How Information Systems Came to Rule the World

This book offers a fresh perspective on information systems, a field of study and practice currently undergoing substantial upheaval, even as it expands rapidly and widely with new technologies and applications.

Mapping the field as it has developed, the author firmly establishes the under-recognized importance of the field, and grounds it firmly in the subject's history. He argues against the view of enthusiasts who believe that the field has somehow moved "beyond information systems" to something more exotic and offers a short and compelling manifesto on behalf of the field and its future.

Offering a comprehensive insight into the significance of the information systems field, this book will appeal primarily to scholars and practitioners working in information systems, management, communication studies, technology studies, and related areas.

E. Burton Swanson is Research Professor of Information Systems at UCLA's Anderson School, USA.

Routledge Focus on IT & Society
Series Editor: Harmeet Sawhney, Indiana University, Bloomington

Routledge Focus on IT & Society provides a showcase for cutting-edge scholarship on social aspects of technology and media. This truly interdisciplinary series offers a space for the emergence of offbeat ideas, works from different perspectives that open a window in the mind, methodological agnosticism, flexibility that operates above barriers to interdisciplinary work, and a focus on conceptual payoff of lasting value.

How Information Systems Came to Rule the World
And Other Essays
E. Burton Swanson

How Information Systems Came to Rule the World

And Other Essays

E. Burton Swanson

Routledge
Taylor & Francis Group

NEW YORK AND LONDON

First published 2022
by Routledge
605 Third Avenue, New York, NY 10158

and by Routledge
2 Park Square, Milton Park, Abingdon, Oxon, OX14 4RN

Routledge is an imprint of the Taylor & Francis Group, an informa business

© 2022 E. Burton Swanson

Library of Congress Cataloging-in-Publication Data
A catalog record for this title has been requested

ISBN: 978-1-032-15336-0 (hbk)
ISBN: 978-1-032-17229-3 (pbk)
ISBN: 978-1-003-25234-4 (ebk)

DOI: 10.4324/9781003252344

Typeset in Times New Roman
by codeMantra

For Cheryl

Contents

Figures

Tables

Preface

Information systems are today ubiquitous in business and organizations of every kind, but in their modern form they are a relatively recent phenomenon arising with the advent of digital computing. While they are now everywhere with us, information systems are also widely misunderstood and underappreciated, even by those whose work and leisure revolves around and depends heavily on them. And yes, even by many information systems professionals themselves.

In this book, addressed to a broad audience of professionals, practitioners, and academics alike, I attempt to bring a better understanding of what information systems are fundamentally all about. I present an updated collection of essays composed mostly over the last decade, each addressing a question I believe to be interesting and important. I admit that this undertaking has been rather self-indulgent on my part, and that the result is a somewhat idiosyncratic composition. I hope it is nevertheless an interesting and provocative read for those like myself who have struggled to grasp and assess the IS phenomenon beyond the ever-continuing excitements of the moment.

I have been at this work a rather long time. My own career in IS began in the summer of 1964 when I returned from a leave at IBM to work in developing applications to manage manufacturing operations in San Jose, California. This was a heady time for IBM, as its third-generation System/360 product line was just in new release and its internal applications across manufacturing locations were planned to be recoded under a common architecture. I was fortunate to be involved in this, and I soaked up the opportunity to learn as much as I could along the way. As my interests developed, in 1968, I began doctoral studies in information science in the business school at UC Berkeley, working part time with IBM.

Upon completion of my studies, I left IBM and worked two years as a guest scientist in Germany before taking an academic position at the Graduate School of Management, UCLA, where many years later I remain today, still curious about IS and where it is going as we look ahead. This book represents an attempt to speak to this.

Whereas years ago there were relatively few IS researchers, there are many around the world today. What now occupies them? A few years ago, the program of the 2018 International Conference on Information Systems (ICIS) listed its contributions under the following track topics (number of sessions in parentheses): bridging the internet of people, data, and things (2); innovation, entrepreneurship, and digital transformation (3); human behavior and IS (4); societal impact of IS and the future of work (1); data science and predictive analytics (4); sharing economy and crowd markets (3); IS in healthcare (4); IS for a green and sustainable world (2); governance, strategy, and value of IS (2); design science (2); practice-oriented, demand-driven IS (2); service science (1); business, data, and process modeling (1); and other (3). At this same time, the leading journal, *MIS Quarterly*, issued a call for a special issue on Next Generation IS Theories, to advance the field in the light of "fundamental changes in the field's core phenomena," where beyond the well-studied world of "distinct users employing clearly distinguishable systems, in bounded contexts, ... IT has become increasingly intelligent, interconnected, and infused through all our contexts, and ... new systems, industries, platforms, and ecosystems have arisen that were previously unimaginable." A workshop to develop papers for the special issue was held at ICIS 2018.[1]

How is it that we have arrived with IT over the course of my career at what was "previously unimaginable" and how should we understand where we now are? This book's essays address this broad question.

Chapter 1, Quo Vadis, Information Systems?, sets the stage for the chapters that follow. It introduces the concept of an information system as a computer-based system for providing information to an organization to help guide its actions, and reviews familiar IS types as background for the general reader. It then considers whether recent developments such as those in platforms, digital transformation, and big data and predictive analytics, among others, have taken us beyond information systems in some fundamental way, and argues, to the contrary, that IS underpins all of this and is more important than ever.

Chapter 2, Why Do Firms Have Information Systems?, addresses the organizational foundations for information systems. It considers how IS helps guide organizational actions, in particular. It explores the idea that, while the need to know by both firms and people is rooted in their own actions, it is largely as these actions pertain to their interactions. A multilevel interaction perspective of IS in organizations is presented. This notion of interaction is a foundational concept of the book.

Chapter 3, What Information Is Provided by Information Systems?, concerns itself with the sticky issue of how we should think about information from the IS viewpoint. Building from a concept rooted in human communication, it suggests that information be conceived as purported facts given and taken, and inferences drawn and established, by participants in an organizational situation. It argues further that in today's systems such information is not stored as such, as in a database, but rather arises and is maintained in an open interaction network of people and machines.

Chapter 4, Why Is Everyone Now an Information System User?, examines the extension of IS beyond the organization to engage the individual person as a consumer and in other roles and activities apart from traditional system use as an employee. It suggests that there are four basic forms of interaction among individuals—informational, transactional, cooperational, and social—that account for everyday personal use of laptops, tablets, and smart phones in accessing interaction networks facilitated by information systems.

Chapter 5, How Did Information Systems Come to Rule the World?, provides a historical interpretation of how IS has come to under-recognized global prominence. It suggests that IS has come to "rule" the world not as kings or other authorities, but rather more literally, through actual rules they embody, such that they dictate how much of everyday life, as it relates to individuals and organizations, takes place around the globe. They rule mostly without drama as infrastructure. The everyday life referred to involves our individual interactions with organizations and with each other, and, in particular, the transactions we necessarily engage in as we go about our personal and working lives.

Chapter 6, How Do Human Practices Change with Information Systems?, speaks about how technological change with IS should be understood, so that we can not only better see the past and present, but also the future in the making. Taking a broad view, it proposes that technology be understood not as physical stuff, as is common today, but rather as capabilities achieved through the fusion of devices and routines as we collectively seek to advance human practices. It examines the multiple ways in which change in technology and practices comes about in the case of information systems, in particular.

Chapter 7, How Can Information Systems Make a Better World?, ventures to answer the tough questions of what the future might hold for IS should we do our best to make the world a better place. Drawing from discussions at a recent workshop, it addresses the dilemmas posed by the "enemies" of the systems approach commonly undertaken and offers a set of new guidelines for our future work.

Together, these seven chapters aim to give the reader something of a fresh perspective on information systems. For the practitioner, I hope it gives insight and a new sense of the importance of information systems to his or her profession, whether it be in the design and development of new systems or in making the best use of them in whatever context, not only in business firms but also in organizations of every kind. Today, especially among the professions, everyone is likely a user of information systems.[2]

For the academic, and especially for the reflective student of systems and organizations, in whatever related field, I hope it provokes new thinking about the research he or she might and should undertake as we engage the challenges posed by information systems in the years ahead.[3]

Reference

Burton-Jones, A., Butler, B., Scott, S. and Xu, S.X., (2021). Next-generation information systems theorizing: A call to action. *MIS Quarterly, 45*(1), 301–314.

Notes

1 The special issue of MISQ is now published (see Burton-Jones et al., 2021, for the introductory editorial). ICIS is the leading research conference in the IS field. The leading academic association is the Association for Information Systems (AIS), with some 4,500 members worldwide. AIS is notable for its international organization, which features three regions (The Americas; Europe, Middle East, and Africa; Asia and the Pacific), and a leadership and conference system that rotates among them. AIS also maintains a digital library that provides rich access to the important publications in the field. See https://www.aisnet.org.

2 Graduate education in the professional schools, such as those at my own university, UCLA, typically reflects the importance of information systems, not only to management, and to information science, computer science, and communication professionals, but further to engineering, medicine, public health, education, law, and architecture, among others, all of which have practices underpinned by IS. We touch on several of these in the chapters ahead.

3 While information systems is now a well-established field of study in its own right, many of its contributing scholars come from related fields. It has always been so. In reflecting on my own doctoral student years, prior to the field's establishment, important contributions were made from management scientists represented by The Institute of Management Sciences (TIMS), now The Institute for Operations Research and Management Science (INFORMS), from organization scientists represented by the Academy of Management (AoM), and from computer scientists represented by the Association for Computing Machinery (ACM). Each of these associations, to their credit, made room for the emergence of IS as a new field of study. They and their members continue to be important contributors to the now established field, as do many other related fields, associations, and their members. Internationally, the International Federation for Information Processing (IFIP) has also played an important role. Among its technical committees, TC8, on Information Systems, has long been composed of working groups helpful to the development of the field. I have myself been long involved with WG8.2 on The Interaction of Information Systems and the Organization, and WG8.6 on Transfer and Diffusion of Information Technology.

Acknowledgements

Chapter 1 draws in part from an entry on information systems published in the *Encyclopedia of Library and Information Sciences*, third edition, 2010. It is otherwise original to the book.

Chapter 2 adapts, updates, and expands upon the paper, "Why do firms have information systems," presented at the Americas Conference on Information Systems, held August 9–12, 2007, in Keystone, CO.

Chapter 3 is adapted from the paper, "Organizational information in the cloud of interaction," presented at the Americas Conference on Information Systems, held online, August 10–14, 2020.

Chapter 4 is adapted from the article, "Who learns what from the new human-computer interaction," presented at the Americas Conference on Information Systems, held August 9–12, 2012, in Seattle, WA, and subsequently published in *The Data Base for Advances in Information Systems*, 44 (1), 9–17, 2013.

Chapter 5 is adapted from the article, "How information systems came to rule the world: Reflections on the information systems field," published in *The Information Society*, 36 (4), 1–15, 2020.

Chapters 6 expands upon a paper presented at the 4th International Workshop on The Changing Nature of Work (CNoW) held December 11, 2016, in Dublin, and draws too from a paper presented at the Academy of Management Meeting, held August 5–9, 2016, in Anaheim, CA, and further developed and published as "Technology as routine capability," *MIS Quarterly*, 43 (3), 1007–1024, 2019.

Chapter 7 is original to the book. It is inspired in part by discussions at the ad hoc Workshop on Information Systems Research and Development (WISRD) held August 19, 2019, in Santa Barbara, CA.

Acknowledgements

1 Quo Vadis, Information Systems?

What are information systems and where are we going with them? In this introductory chapter, I briefly summarize what information systems are all about, as I see it, and how this has come to be, and what the issues are as we face the future. Much of what follows will be familiar to scholars and practitioners in the IS field. But, for the general reader, I hope this background will be helpful. Subsequent chapters address issues introduced here.

As we shall view it, an information system is commonly a computer-based system for providing information to an organization to help guide its actions. To say that an IS is computer-based is not to say that it is computer-confined, as will be emphasized. To say that it provides information to an organization leaves unsaid for the moment how it does so. To say that it helps to guide organizational actions does explicate the rationale for its existence.[1]

Typically, an information system features people working interactively with computers to accomplish a particular task. As seen here, both people and machines are informed through these interactions. Often, the information provided serves to coordinate workers' specialized but necessarily collective efforts. Where associated decisions are routine and highly structured, they are sometimes relegated to the machine. Where not, they are often left to the workers. The varieties of information systems are many, reflecting the diversity of organizations and tasks to be accomplished. A typical large business firm has information systems to support its accounting and finance, operations, supply chain management, sales and marketing, customer service, human resource management, and research and development. But information systems are found everywhere, in organizations of all kinds and sizes, public as well as private.[2]

The technical heart of any information system, that which is computer-driven, consists of application code and associated data.

DOI: 10.4324/9781003252344-1

The term "data processing" describes in a nutshell what the computer does when the code is executed. But the larger notion of an IS incorporates not only such data processing, but also the acquisition of the data, and the employment of outputs in whatever form to guide or effect actions by either humans or machines. The acquisition and employment processes may be automated or not, according to the design of the IS.[3]

In what follows, I elaborate. I first discuss the origins of information systems in the field of practice. I then review the basic types of information systems as commonly found. I then touch upon their recent extensions, which are varied and many. Finally, I discuss their futures and set the stage for the chapters to follow.

Origins

Modern information systems emerged with the rise and spread of digital computing in the 1950s, although punched card tabulating equipment was in use for data processing in organizations before then. The stored-program computer itself was initially viewed as a high-powered calculating device, suitable primarily for numerical and other sophisticated analyses. Such "scientific computing" was distinguished from what was termed "electronic data processing" (EDP), which emerged about the same time to support the more prosaic work of business, such as accounting. In the 1960s, computers came to be designed and marketed specifically for business purposes, eventually displacing the tabulating equipment. Notably, a high-level programming language for business applications, Common Business-Oriented Language (COBOL), was also developed, which emphasized data and file structures, and deemphasized the computational features found in FORmula TRANslation (FORTRAN), the language most commonly used in scientific computing. COBOL ultimately became the most widely used programming language for the development of application software for information systems on mainframe computers. As much of this code remains in use, the language persists even today.[4]

Beyond business-oriented application software, the emergence of data base technology in the late 1960s was central to the rapid rise and spread of large-scale information systems among firms. A data base is an organized collection of related data files. A data base management system (DBMS) is system software that enables data bases to be managed as integrated wholes, where relationships among files are clearly delineated. With a DBMS, data can be defined via a data dictionary and managed separately from the different software which access it. The articulation of the relational data model as a foundation for data bases spurred the development of relational data bases in the 1970s, which came to dominate the field. Today, Oracle provides the leading relational data base software for medium to large firms, while Microsoft's Access is well established among small businesses.[5]

Together, application software and a related data base have come to form the digital content around which any modern information system is now built. Typically, the application software incorporates the "business rules" to be followed, while the data base incorporates the "business facts" that shape the data processing, for instance, in processing a business payroll, or in selling seats to a concert, or in managing the circulation of a library's holdings, or in almost any other endeavor in which carefully informed organizational actions are routinely taken. While the business facts and data base will typically be specific to the enterprise, the business rules and application software may be either specific or generic, i.e., commonly used, as with accounting systems that incorporate professionally mandated rules and principles. Where the business rules and application software is specific to the organization, it may underpin the unique capabilities of the enterprise, in which case it may be strategic. Today, people in a wide variety of occupations and in organizations large and small are likely to work interactively with information systems to accomplish much of their work. Through networks and the Web and Internet, in particular, and through the use of laptops and mobile devices they engage in this "human-computer interaction" (HCI) from wherever they happen to be and at whatever times they choose or are called upon to be available.[6]

Types

Information systems come in a wide variety, reflecting the diversity in the organizations that employ them. Among business firms, some information systems will be characteristic of the industry, in particular, as with process control systems in chemical and refining enterprises, or electronic funds transfer (EFT) systems in banks and other financial services firms. However, certain basic types are found in enterprises of all kinds, reflecting both their historical origins based in then-new technologies and the nature of organization itself. These include transaction-processing systems, management information systems (MIS), decision support systems (DSS), group support systems, and enterprise systems. These are not pure types; actual systems may combine features of two or more basic types.[7]

Transaction-Processing Systems

Transaction-processing systems support an enterprise in its transactions with others, such as customers, suppliers, and employees. Every business transaction involves an exchange of goods, services, money, and information in some combination between the parties. Transaction-processing systems exist to ensure the integrity of these transactions. In today's world, each time a consumer makes a purchase with a credit card, withdraws cash from an account, or books an airline ticket, the consumer likely engages the other party's transaction-processing systems. Increasingly, a consumer does this directly, by positioning a bank card or smart phone at a point-of-sale (POS) device or employing an automated teller machine (ATM) or initiating a purchase from the Web.[8]

Beyond their primary function, transaction-processing systems also enable a business to coordinate its internal operations among units, especially in the making of goods, where parts are withdrawn from inventory and a manufactured item is assembled in a series of operations, and the final product eventually distributed from one location to another, for instance. Here and elsewhere, transaction-processing systems are basically event-driven, and are often engaged to authorize formal actions, such as accepting a customer order or authorizing a credit purchase. The business rules for such data processing may be quite sophisticated, as in credit authorization which incorporates rules aimed at fraud detection, for instance. The data pertaining to these events will ultimately serve to update a data base that is typically drawn upon in processing and is relied upon to give the current status of the organization's affairs. Where the data base is immediately updated as events happen, the system is said to operate in "real time." In the case of firms, basic transaction data will further feed the accounting systems that provide a formal financial picture of the ongoing business.

Where firms do business with each other, for instance, within a supply chain, their transaction-processing systems are also sometimes tied together by means of an interorganizational system that enables them to communicate directly with each other. For such machine-to-machine communication, this necessitates resolution of disparities in how the data themselves are defined by the communicating parties. The interorganizational system may be based on electronic data interchange (EDI) arrangements or increasingly on eXtensible Markup Language (XML) standards for exchange over the Web. The concept of Web services envisions a world of business services and firm transactions seamlessly tied together via standards for business data of all kinds.[9]

Management Information Systems

Management information systems support an organization's hierarchical structure and are targeted to management at all levels. MIS aim to support every manager's need to know within his or her scope of responsibility, typically by extracting important performance information from data gathered from the organization's transaction-processing and operational systems and presenting it efficiently in tabular or graphical form. The concept of an MIS emerged in the 1960s and signaled an important transition in information systems, from traditional EDP to systems that served more sophisticated purposes. In the United States, both practitioners and educators embraced the MIS concept and many business schools originated programs of study under this banner. Today the term continues to be widely used, although the more generic term "information systems" has become more common.[10]

Executive information systems (EIS) were founded in the 1980s as a new form of MIS aimed at top management. The early EIS featured access to news external to the business, in addition to traditional performance metrics, and further employed new graphics and communications technologies. Most recently, executive support systems have been developed which provide a personalized Web page and "executive dashboard" of up-to-the-minute information with which the manager is to engage and steer the enterprise. These systems are now also popularly referred to as "business intelligence systems," reflecting the sophisticated analytics that may lie behind the dashboard metrics.[11]

Decision Support Systems

Decision support systems emerged in the 1970s as interactive systems that supported managers and other "knowledge workers" in tasks that were semi-structured, where decisions could be aided by analytical computer-based means. These systems shifted the original MIS focus from information to decisions. Early DSS featured innovative HCI employing graphics, formal models, and heuristics or algorithms as means of support. A pioneering example was IBM's Geodata Analysis and Display Systems (GADS), which supported organizational decisions related to urban geography, such as arranging police beats and assigning school district boundaries. Today, the concept of geographical information systems (GIS) continues in this tradition as a major area of application supported by new technologies such as remote sensing, geographical positioning, graphical analytics, and visualization. The firm ESRI is the leading provider of GIS software.[12]

The concept of group decision support systems (GDSS) extended the basic DSS concept in the 1980s. Substantial research led to the development of decision rooms equipped with systems that facilitated complex, interactive group decision-making in a particular location. The early focus was typically on largely unstructured problems, with tools provided to support collective brain-storming and idea evaluation, for instance, while further capturing a record of the group meeting. With advances in communications technologies, the GDSS concept soon evolved into one that supported group work more broadly, where group members could be at multiple locations and could also meet asynchronously as needed.[13]

Group Support Systems

Beyond the informational and decisional needs of managers, it is well understood that communication and cooperation more broadly in the organization is required to coordinate the work undertaken within and across units. Certain of this communication and cooperation can be built into the work systems themselves; however, other organizational means such as cross-functional teams can also facilitate lateral communication, cooperation, and coordination, thus moderating the burden on the management hierarchy. Not surprisingly, given the ubiquity of group work in organizations, a wide variety of systems have in recent years been originated to support group work, in particular.[14]

The concept of computer-supported collaborative work (CSCW) originated in the 1980s to characterize designs for computer-enabled group work, understood to require substantial communication and coordination, typically over time and across locations. Lotus Notes exemplified the software then deployed in these new systems and remains in use today. Current groupware in support of group work is diverse and includes, e.g., that which provides for electronic meetings, electronic mail and messaging, calendar management, project and document management, knowledge sharing, workflow management, and collaborative design. Today, group work can also be organized and conducted on the Web, making use of a commercially available service.[15]

Enterprise Systems

Enterprise systems emerged in the 1990s with the rise of enterprise resource planning (ERP), a concept for integrating the major functional systems of the enterprise, in particular, the organization's financial, human resource, and operational systems around a common data base. The principal means of integration was typically a software package provided by a vendor such as SAP or Oracle. Firms sought to replace their older and disparate home-grown legacy systems, which required high maintenance, with standard off-the-shelf software that promised an integrated solution to relieve them of this burden. ERP basically incorporated the firm's major transaction-processing and operational systems. Its adoption was further given a large boost by concerns related to the millennium bug and the threat it posed at the time to vulnerable legacy systems. Today, most large firms have adopted and implemented ERP in the form of packaged software provided by one or more leading suppliers.[16]

A second type of enterprise system termed "customer relationship management" (CRM) has also become popular, focusing on the "front office" of a firm, beyond the already heavily computerized "back office." A central CRM aim is to provide the firm with a "unified view" of its customers, who might otherwise engage in separate transactions with different business units, each in the absence of full customer information. Just as it promises better customer service, CRM also typically supports a firm's sales force and enables it to be better managed. Still another CRM aim is to help the firm assess the profitability of its different customer segments, in the interest of focusing marketing and customer retention initiatives on achieving higher overall profits.[17]

Extensions

One of the most remarkable aspects of information systems pertains to what I will term their *extensions*. Whereas in their early days, information systems were largely confined to large organizations that could afford them, and were something of a mystery to many people, today they are found pretty much everywhere, and familiar to most workers. Moreover, they have extended their worldly reach in multiple ways, several of which stretch our thinking of what they are now about.

Consider the following laundry list of extensional directions over just a few decades: from large to medium and small businesses; from firms to consumers; from big to small computers; from stationarity to mobility; from isolated to networked operation; from centralized to decentralized computing; from human to artificial intelligence; from small to big data; from business function to business enterprise; from established industry to new industry; from the developed world to the developing world.

All this is rather a lot! And where have we arrived? As one indicator, consider that the five most valuable business firms among the Fortune 500 in terms of their market capitalization were as of May 21, 2018: Apple ($921b), Amazon.com ($765b), Alphabet ($750b), Microsoft ($746b), and Facebook ($531b). All of these giants are deeply engaged with information systems, both as providers of products and services to others with information systems and as users of systems themselves. While it might not be immediately apparent, the value they all now represent is largely a value achieved through their own information systems and those of their customers.[18]

Various extensions to information systems have challenged our notion of what these systems are all about and some scholars have suggested that we have moved "beyond information systems." A good example is in digital entertainment, as with Netflix. Certainly, digital representation and the streaming of video in this case goes beyond information systems of the classic types discussed above. But what is less well understood is that such a business is built fundamentally on information systems that allow for the digital content to be assembled, characterized, organized, stored, and accessed on request, supported by delivery, subscription, and payment systems, among others. There is no "beyond information systems" here. Quite the contrary. Even the current intense interest in algorithms, as exemplified with Netflix's recommender system, is but the latest variation in the application code needed to support an enterprise. Which is why I refer to these new developments as extensions.[19]

As another example, a current popular concept with managers is "digital transformation" of the business. Exactly what this amounts to is often rather vague. According to one research report,

> The world is going through a kind of digital transformation as everything—customers and equipment alike—becomes connected. The connected world creates a digital imperative

for companies. They must succeed in creating transformation through technology, or they'll face destruction at the hands of their competitors that do.

However such a warning is interpreted, it is certainly true that digital content in various new forms is leading to upheavals among industries and their products and services (healthcare offers a particularly timely example). Most of this, I suggest, is at bottom due to the transformational extension of information systems that provide for this new content and allow for it to be deployed.[20]

Today, much of this transformational extension is taking place by means of *digital business platforms* that facilitate business innovation by providing a host service of importance that also capitalizes on economies of scale. Amazon provides a particularly dominant example, serving both producers and consumers of both physical and digital goods through its digital business transaction platform, as well as its warehousing and logistics system. It further provides cloud computing capacity through its complementary Web Services business. Amazon's rise has of course been transformative to businesses and the economy. Most importantly, from the viewpoint of this book, Amazon as an enterprise is built entirely on IS foundations.[21]

Still another transformational extension, made possible through platforms, is the development of *digital business ecosystems* in which firms partner in delivering customized product-service bundles to customers (Jacobides, 2019). Such ecosystems are now an important feature of our modern economy, and they typically are built through data sharing arrangements among partners.

In sum, information systems are not going away. Rather they are growing in number and kind, and elaborating in their forms. Importantly, so too are the organizations that employ them. With the arrival of digital business ecosystems, one is tempted to suggest that an information system maxim has been demonstrated: *organizational form follows information system function.*[22] Has this perhaps always been the case, over the history of IS? Regardless of one's view on this, the challenge for IS scholars is to better articulate both the fundamentals of information systems and the many new forms these systems can assume as they are built from knowledge of these fundamentals.

Futures

What are the likely futures for information systems? This we explore in the chapters that follow. Here, to set the stage, we can extrapolate briefly from current developments. We can envision a world in which

IT is even more ubiquitous: (i) bots and robots come to dominate an increasing portion of everyday life; (ii) monitoring of physical environments and human activities is increasingly prevalent around the globe; (iii) organizations increasingly thrive or not according to the knowledge gained from their systems, more than from their people; and (iv) humans are increasingly sustained or not by their access to wealth, as much as by their access to traditional employment.[23]

With regard to bots and robots, the question is for whom will they work? While some may work for people in the home, most will likely do specialized work for organizations. They will not be autonomous. Rather, they will be tethered to the organization as agential extensions to its operational systems. Similarly, monitoring of physical environments and human activities will not take place as independent activity. Rather, it will be hosted by organizations that gather and accumulate its information for the systems that then make use of it.[24]

Concerning organizations, while human capital will rightly continue to receive great attention as to its importance, I suggest that most enterprises will nevertheless succeed or not according to the knowledge gained from their information systems. In fact, the value of human capital will often be in expert service to the organization's systems.[25]

Perhaps the most important consequence of our increasingly automated future pertains to the future of human work. Here I am concerned with the prospects for paid work to sustain human populations where the returns to capital progressively swamp the returns to labor. I suspect that we will need to give many people better access to created wealth beyond paying jobs, perhaps in the form of social dividends granted, as with basic income proposals. How we might best do this I do not know at present. But whatever is undertaken, we can be certain that information systems will undergird the organized endeavor.[26]

As should be apparent from these brief remarks, I see the future of information systems tightly bound up with the future of organizations. Today, all organizations require IS to manage their related activities, and the systems themselves are organizationally hosted even when not contained as such within the single enterprise. The relationship of IS to organizations is, however, increasingly complex, both in theory and in practice. How things will unfold over the next decades is quite uncertain. How we should understand information systems in organizational context as we look to the future for both is a major theme of this book.

Reflections

In the chapters to follow, I reflect further on what information systems are about and where we are going with them. Each chapter poses a question. Chapter 2 asks, Why do firms and other organizations have information systems? This would seem to be a rather easy question to answer on the face of it. But seeking more than a superficial grasp of the information systems phenomenon and what lies behind it, I attempt answering the question in some depth. What emerges is a multi-level interaction perspective on the role of IS in organizations. Succeeding chapters pose related questions. While they are ordered to be read in succession, they may also be read independently.

Notes

1 The concept is similar to that advanced by Langefors (1977), among others. The term is also sometimes used in information science to refer to information retrieval systems based more on documents than on data, an application domain familiar to libraries, in particular. It is sometimes also used very generally and informally, without reference to either computers or organizations. People sometimes refer to their own personal information systems, for instance. Here I take an organizational perspective, which has its origins in business but applies to organizations of every kind.

2 Simon (1960) identifies "programmed decisions" as those which by their structure can be suitably executed by machine. The term is apt. All computer programs make decisions in execution, mostly many very small and incremental ones that then accumulate, according to the data processed, in whatever application context. Computers are inherently decision-making machines. Incorporated in information systems, they make the programmed decisions specific to the organizations they serve.

3 Swanson (1991) incorporates this interpretation of information systems in an attempt to explain how they are anchored and woven into organizations. Today, much of this weaving is substantially automated. For instance, with Big Data, much is acquired by sensors or as a by-product of human online activity, giving rise to "data exhaust" which can be subsequently mined to extract organizational value in uses beyond those that generated it. See George et al. (2014) and Constantiou and Kallinikos (2015) on Big Data. Subsequent employment of the data as processed may also be automated and woven across organizations, not only within them, as with platforms in e-commerce.

4 Canning (1956) provided an important early work on electronic data processing (EDP).

5 Date (1981) offered an early authoritative work on data bases. Codd (1972) introduced the concept of the relational data base. Historically, computer science was slow to recognize the importance of data

in comparison to computation. Today, with Big Data, the tables have been turned. Now it is data science that stokes the interests of those putting computers to use. See, e.g., Dhar (2013).

6 The concept of "strategic information systems" became an important one in the field. Porter and Millar (1985), in a highly influential article, explained how information systems can provide for competitive advantage. See too Porter's (1985) classic work on competitive strategy more broadly.

7 We note that where information systems are specific to an industry, they are likely closely linked to its core technology, and are all the more indispensable, and often strategic in their specifics. Such technology is often overseen by a chief technology officer (CTO), who works closely with the chief information officer (CIO). The history of airline reservation systems provides a classic example (Hopper, 1990), which also illustrates how systems can become so important that they become businesses in themselves.

8 Zwass (1998), Chapter 9, provides a good discussion of transaction processing systems internal to the firm.

9 Johnston and Vitale (1988) discuss interorganizational systems and how they may be used to competitive advantage. Hagel and Brown (2001) provide an introduction to web services and their strategic implications.

10 Dickson's (1981) article on management information systems remains a significant source for scholars interested in the history of the field. See also Society of Management Information Systems (1970) for its founding description of MIS. Founded in 1968, SMIS became the leading practitioner association for IS executives in the United States and Canada, and renamed itself Society for Information Management (SIM) in 1982. Today it has some 5,000 members among 40 chapters.

11 Watson et al. (1981) provided an authoritative introduction to executive information systems at the time. Gray (2006), Chapter 8, discusses business intelligence systems.

12 Keen and Scott Morton (1978) provided an early introduction to decision support systems. Sprague and Carlson (1982) include a useful description and discussion of GADS (pp. 41–54).

13 Dennis et al. (1988) discuss technology for the support of group meetings at this early time.

14 Galbraith (1973) discusses the importance of lateral communications to management of the hierarchy. The common assumption at the time was that this was a human communication problem, not pertinent to IS, whereas vertical communication, being managerial, lent itself to support through MIS. This assumption proved to be short-sighted with the rise of computer support of communications at and across every level of the enterprise.

15 Grudin (1994) provides a helpful history of CSCW. Another group support concept that originated in the 1990s was that of knowledge management. Alavi and Leidner (2001) examine the concept and its implications for research.

16 Davenport (1998) offered an early article on enterprise systems.

17 See Winer (2001) on CRM from a marketing perspective. A third type of enterprise system is that of supply chain management (SCM). Lee (2004) provides a managerial perspective on the challenges of modern supply chains.

18 The market capitalization numbers are taken from the following link: fortune.com/2018/05/21/fortune-500-most-valuable-companies-2018/. The values represented should of course not be equated to social value. Rather they reflect the future profits the companies are expected to achieve. The numbers fluctuate according to expectations for the companies and the equity market as a whole. Zhu and Furr (2016) discuss the rise of Amazon, Google, and Apple from their product origins to their current platform forms.

19 See Gomez-Uribe and Hunt (2016) on Netflix's recommender system. Adhikari et al. (2012) examine the Netflix architecture for content delivery.

20 The quote is taken from Fitzgerald et al. (2014, p. 4). Comparable warnings that companies need to get with the latest in new information technology or face disastrous consequences if they do not have been with us from the beginning. Agarwal et al. (2010) provide a substantive discussion of the digital transformation of healthcare.

21 Eisenmann et al. (2011) elaborates on platforms and business strategy: "A platform-mediated network is comprised of users whose interactions are subject to network effects, along with one or more intermediaries who organize a platform that facilitates users' interactions. The platform encompasses the set of components and rules employed in common in most interactions between network users. Components include hardware, software, and service modules, along with an architecture that specifies how they fit together. Rules encompass information visible to network participants that is used to coordinate their activities. In particular, rules include standards that ensure compatibility between different components; protocols that govern information exchange; policies that constrain network user behavior; and contracts that specify terms of trade and the rights and responsibilities of network participants" (p. 1272). Note that the rules governing platform participation would be incorporated in the platform's IS. Ritala et al. (2014) include a case study of Amazon's business model.

22 The notion that form follows function is a design principle from late 19th and early 20th century architecture and industrial design, asserting that the shape of a building or object should reflect its intended function or purpose. Applied to organizations and information systems, the maxim suggests that any organizational form should reflect the functions of the information systems that enable it to do business with its customers, as with the new digital business ecosystems. This is admittedly something of a conceptual leap. We make this suggestion here only to provoke thinking on the subject.

23 Perhaps the most expansive vision of the future is that of the Fourth Industrial Revolution (Schwab, 2017), in which multiple technologies, including artificial intelligence, robotics, internet of things (IoT), quantum computing, fifth generation wireless, biotechnology,

nanotechnology, and 3D printing, converge to completely transform the world economy.

24 A good example of specialized work featuring bots and robots in increasing roles is that of customer order fulfillment, a sophisticated process in electronic commerce, in particular, as with Amazon. Transactional bots are frequently employed in order taking, while warehouse robots help move the goods around as needed to deliver on the order. Interestingly, picking items from warehouse shelves has been notoriously hard to automate (Agrawal et al. 2018).

25 Goldin (2016) provides an overview of the economics of human capital, defined as "the stock of productive skills, talents, health and expertise of the labor force" (p. 80). Importantly, investments in human capital and technology are intertwined in something of a race between them. "Technology complements skill and increases the returns to investment in education. Education, in turn, induces more technological change" (pp. 73–74).

26 See Van Parijs and Vanderborght (2017) for a thorough examination of basic income proposals. Piketty (2014), in an influential treatise, documents the increased concentration of wealth and suggests that it is not tenable in maintaining democratic values. Brynjolfsson and McAfee (2014) examine the respective returns to labor and capital in an increasingly automated economy (pp. 142–146) and question whether the historical balance will be maintained, arguing that basic income proposals should be in the mix in the search for equitable wealth distribution.

References

Adhikari, V. K., Guo, Y., Hao, F., Varvello, M., Hilt, V., Steiner, M. and Zhang, Z. L. (2012). Unreeling Netflix: Understanding and improving multi-cdn movie delivery. *INFOCOM, 2012 Proceedings IEEE*, 1620–1628.

Agarwal, R., Gao, G., DesRoches, C. and Jha, A. K. (2010). Research commentary—The digital transformation of healthcare: Current status and the road ahead. *Information Systems Research, 21*(4), 796–809.

Agrawal, A., Gans, J. and Goldfarb, A. (2018). *Prediction Machines.* Boston, MA: Harvard Business Review Press.

Alavi, M. and Leidner, D. E. (2001). Knowledge management and knowledge management systems: Conceptual foundations and research issues. *MIS Quarterly, 25*(1), 107–136.

Brynjolfsson, E. and McAfee, A. (2014). *The Second Machine Age.* New York: W. W. Norton.

Canning, R. (1956). *Electronic Data Processing for Business and Industry.* New York: Wiley.

Codd, E. F. (1972). A relational model of data for large shared banks. *Communications of the ACM, 13*(6), 377–387.

Constantiou, I. D. and Kallinikos, J. (2015). New games, new rules: Big data and the changing context of strategy. *Journal of Information Technology, 30*(1), 44–57.

Date, C. J. (1981). *An Introduction to Data Base Systems*, 3rd ed. Reading, MA: Addison-Wesley.

Davenport, T. H. (1998). Putting the enterprise into enterprise systems. *Harvard Business Review, 76*(4), 121–131.

Dennis, A. R., George, J. F., Jessup, L. M., Nunamaker, J. F., Jr. and Vogel, D. R. (1988). Information technology to support meetings. *MIS Quarterly, 12*(4), 591–624.

Dhar, V. (2013). Data science and prediction. *Communications of the ACM, 56*(12), 64–73.

Dickson, G. W. (1981). Management information systems: Evolution and status. *Advances in Computing, 20*, 1–37.

Eisenmann, T., Parker, G. and Van Alstyne, M. (2011). Platform envelopment. *Strategic Management Journal, 32*(12), 1270–1285.

Fitzgerald, M., Kruschwitz, N., Bonnet, D. and Welch, M. (2014). Embracing digital technology: A new strategic imperative. *Sloan Management Review, 55*(2), 1.

Galbraith, J. R. (1973). *Designing Complex Organizations*. Reading, MA: Addison-Wesley.

George, G., Haas, M. R. and Pentland, A. (2014). Big data and management. *Academy of Management Journal, 57*(2), 321–326.

Goldin, C. D. (2016). Human capital. In Diebolt, C. and Haupert, M. (eds.), *Handbook of Cliometrics*. Heidelberg, Germany: Springer Verlag, 55–86.

Gomez-Uribe, C. A. and Hunt, N. (2016). The Netflix recommender system: Algorithms, business value, and innovation. *ACM Transactions on Management Information Systems (TMIS), 6*(4), 13.

Gray, P. (2006). *Manager's Guide to Making Decisions about Information Systems*. New York: Wiley.

Grudin, J. (1994). Computer-supported cooperative work: Its history and participation. *IEEE Computing, 27*(5), 19–26.

Hagel, J., III and Brown, J. S. (2001). Your next IT strategy. *Harvard Business Review, 79*(10), 105–113.

Hopper, M. D. (1990). Rattling SABRE-new ways to compete on information. *Harvard Business Review, 68*(3), 118–125.

Jacobides, M. G. (2019). In the ecosystem economy, what's your strategy? *Harvard Business Review, 97*(5), 128–137.

Johnston, R. and Vitale, M. J. (1988). Creating competitive advantage with interorganizational information systems. *MIS Quarterly, 12*(2), 153–165.

Keen, P. G. W. and Scott Morton, M. S. (1978). *Decision Support Systems: An Organizational Perspective*. Reading, MA: Addison-Wesley.

Langefors, B. (1977). Information systems theory. *Information Systems, 2*, 207–219.

Lee, H. L. (2004). The triple-A supply chain. *Harvard Business Review*, *82*(10), 102–113.

Piketty, T. (2014). *Capital in the Twenty-first Century*. Cambridge, MA: Belknap Press.

Porter, M. E. (1985). *Competitive Advantage*. New York: The Free Press.

Porter, M. E. and Millar, V. E. (1985). How information gives you competitive advantage. *Harvard Business Review*, *63*(4), 149–160.

Ritala, P., Golnam, A. and Wegmann, A. (2014). Coopetition-based business models: The case of Amazon.com. *Industrial Marketing Management*, *43*(2), 236–249.

Schwab, K. (2017). *The Fourth Industrial Revolution*. New York: Crown.

Simon, H. (1960). *The New Science of Management Decision*. New York: Harper & Row.

Society for Management Information Systems. (1970). What is a management information system? Research report no. 1 Society for Management Information Systems, Chicago, IL.

Sprague, R. H., Jr. and Carlson, E. D. (1982). *Building Effective Decision Support Systems*. Englewood Cliffs, NJ: Prentice-Hall.

Swanson, E. B. (1991). The information loop as a general analytic view. *Information & Management*, *20*, 37–47.

Van Parijs, P. and Vanderborght, Y. (2017). *Basic Income*. Cambridge, MA: Harvard University Press.

Watson, H. J., Rainer, K. and Koh, C. (1991). Executive information systems: A framework for development and a survey of current practice. *MIS Quarterly*, *15*(1), 13–30.

Winer, R. S. (2001). A framework for customer relationship management. *California Management Review*, *43*(4), 89–105.

Zhu, F. and Furr, N. (2016). Products to platforms: Making the leap. *Harvard Business Review*, *94*(4), 72–78.

Zwass, V. (1998). *Foundations of Information Systems*. Boston, MA: Irwin/McGraw-Hill.

2 Why Do Firms Have Information Systems?

Why do firms have information systems (IS)? In this chapter, I ask this seemingly innocuous question to explore some foundations for IS study and practice in a business context. Over the last half century or so, with the rise of digital computing, modern IS of various kinds have emerged and become ubiquitous among firms. But both theorists and practitioners still struggle to describe and explain what has been and is still being accomplished with this transformation. A broad understanding and appreciation of IS beyond those immediately committed to it in their work seemingly remains to be achieved. Here I seek to make a small contribution toward this understanding. While the chapter focuses on business firms, we will see that the answer to our question applies to organizations of all kinds.

To begin then to address our question, let's consider a familiar everyday phenomenon, where downtown shoppers routinely approach and engage a bank's automated teller machines (ATMs), often simply to withdraw needed cash from their accounts. These shoppers will give little or no thought to *how* these ATMs work, much like human tellers would work, in interaction with them, asking for identification and responding to their straightforward requests. Of course, upon reflection, these ATMs, like human tellers, can work only because they have access to the bank's online systems, where information about customers and their accounts resides. Also, for the customer's convenience, ATMs, unlike their human counterparts, have this access 24 × 7. So, a first answer to our question, and a good one insofar as it goes, is that *firms have IS so that they can readily do business with their customers.*[1]

More broadly, consider that in doing all manner of business, such as that just illustrated, firms basically need to know where they are in whatever they do. They need to know, one way or the other, (1) what business they have already done; (2) what business they are now doing; and (3) what business they should do next. With its ATMs, the bank needs to know, for instance, who now requests what, and whether to meet the request, in the light of past actions. More generally, I argue in this chapter that the firm's "need to know" relative to its own *actions*, past, present, and future, fuels its development of IS that become natural to the business, in the sense that they become an inseparable part of the business, as with the IS that support a bank's ATMs. The firm's compelling need for information, in short, is rooted in the streams of actions it takes.

Consider too that firms need to know relative to their own actions, much like people need to know relative to theirs. Neither firms nor people act in isolation, however. Rather, both firms and people *interact* with purposeful others in their respective environments, as illustrated in the case of banks' interactions with their customers, and how well both firms and people interact has much to do with their successes and failures. While the need to know by both firms and people is thus rooted in their own actions, it is largely as these actions pertain to their interactions. This idea is one I want to explore further here.

Interaction Foundations

In what ways are the firm's actions grounded in their interactions? We will see that these interactions take place across three levels: the firm level, the subunit level, and the system level.

Firm-Level Actions and Interactions

In the case of firms, their actions are directed most fundamentally toward interactions in the markets in which they do their business. The firm's exchanges in these markets are marked by binding *transactions* between parties. Hence, the firm's need-to-know centers first of all on its business transactions with its customers, and then too, with its suppliers in what is termed the value chain that meets customer needs. Viewed in microcosm, the firm undertakes a series of related actions to carry out each of its business transactions, coordinated with a similar series of related actions undertaken by its trading partner, and at any one time the firm needs to know which of these actions have been accomplished, which are underway, and which remain to be initiated. Each such business transaction commonly involves not only the exchange of goods or services and money or claims on money but also information pertinent to the exchange between the parties. Much of this information eventually finds its way into the firm's accounting systems.[2]

Elsewhere, a firm also enters into transactional relationships with its employees, with whom it contracts for ongoing work in exchange for wages and certain other "benefits." Over time, the firm's many interactions with its employees involve substantial exchanges of information. Increasingly, these interactions are mediated by a firm's online human resource (HR) systems. So, for instance, an employee might now access an online system to change the contribution made to her non-taxable savings plan, whereas formerly she would have made a request to the payroll department, or she might now access a list of open job positions posted online within the firm, whereas formerly she might have had little awareness of these opportunities.

Similarly, a firm engages in transactional relationships with its owners, who provide it with capital in exchange for shares of the firm's ongoing profits, and with financial creditors who also expect to earn from the firm's returns. These relationships are of course thoroughly grounded in information exchanged between the parties. Indeed, the financial exchanges themselves are executed in informational terms.

Finally, firms also act interact with others in their broader institutional environment, where in doing business they are subject to public acceptance and governmental oversight and regulation. Certain of a firm's actions and interactions are accordingly directed at securing this acceptance and making its other actions known as necessary to external agencies and authorities. Here again, IS are often developed to support the firm's various actions and interactions in this environment, especially where these are ongoing, as with a corporation's periodic issuance of its formal financial reports or most recently in the United States its development of controls and reports in compliance with Sarbanes-Oxley requirements.

In sum, as a socially and legally recognized economic entity, a firm acts and interacts with many others in the environment in which it does business. It acts as a "whole," but not of course in the manner of an individual person. Hardly a human organism with a central nervous system, the firm is rather a kind of social *organization*, one that must be somehow coordinated across its interacting subunits to execute the business processes that enable it to make and deliver on its commitments.[3]

Subunit-Level Actions and Interactions

Historically, the modern firm's task of coordinating across subunits fell largely to its management and to a hierarchical form that allowed for vertical communication within the form, while necessitating frequent lateral communication across it. Subunits were coordinated vertically through their common higher-level management or laterally through established work procedures and special means such as cross-functional teams, liaison roles, and the like. Through the first half of the 20th century, firms developed many varieties of this costly organizational form, which relied almost entirely on human effort to sustain it. In carrying out the firm's tasks, much of this effort was devoted to paperwork originated and shared among interacting parties, often to serve the business customer, as in filling a customer order.[4]

With the rise of modern IT in the latter half of the 20th century, however, and with the emergence of easily accessible online systems, in particular, the firm's internal as well as external interactions have been gradually, but fundamentally, transformed. Today, a substantial amount of the information processing needed to coordinate the firm is typically automated. Still lacking a coordinative central nervous system, the firm now develops computer-based systems to assume an analogous role, flexibly supporting both centralization and decentralization, and thereby both traditional and radically new organizational forms.[5]

Most significantly, with the rise of computing and new reliance on IS, certain of the firm's systems have themselves become interactive "machine agents" in the coordinative process. They are now interactive machine agents, just as its employees are interactive human agents, in the sense that their actions, in their interactions, are undertaken on behalf of the firm and can formally commit the firm in its marketplace transactions and relationships with external others. The result, I argue, has been the gradual and still not widely understood emergence of a third fundamental level of action and interaction.[6]

System-Level Actions and Interactions

To grasp what has happened in firms consider what is commonly now seen when one walks about an office complex. One observes many instances of individuals at their desks working with their desktop or notebook PCs, often tackling their email, or seeking information from the Web, but in many cases accessing online systems, often by means of the firm's intranet. One sees much of what is termed "human-computer interaction" (HCI), where the individual engages the online system to accomplish in tandem with the system and others what is often a coordinated task. The individual collaborates electronically with others in preparing a report, completes an assigned project task, undertakes online training, arranges a face-to-face meeting, orders supplies, or engages in any number of other everyday tasks mediated by online systems. While we might still observe substantial human interaction in the office complex, we are now likely to see HCI in just as much or even greater amounts, reflecting the rise of machine agency in the firm.[7]

Significantly, the new machine-based systems do not typically represent the interests of single organizational subunits. Rather, they often represent the interests of an "end-to-end" *business process*, such as customer order fulfillment, that cuts across organizational subunits and may extend too to suppliers and distributors, especially with e-commerce.[8] They thus form another means with which the firm divides and coordinates certain of its labor, however departmentalized. Too, because the firm's systems typically pass data among themselves, or share a common data base, they must themselves be coordinated or even integrated, much as traditional subunits have had to be coordinated in the firm's interests. It is in this sense that they come to serve as part of the firm's "central nervous system."

A Multi-Level Interaction Perspective

As should now be apparent, I am arguing that in understanding why a firm has IS, one should focus on the multi-level actions and interactions that both constitute and enable the firm to do its business. We identify these as follows:

* The firm's *firm-level actions and interactions* in doing business with external others, and in securing a favorable business environment.
* The firm's internal *subunit-level actions and interactions* that coordinate the business across its subunits.
* The firm's *system-level actions and interactions* among various human agents and machine systems, that serve the above levels of actions and interactions.

Consider that while the first two levels characterize the firm's need for IS, the third level characterizes the way in which IS themselves function to meet this need. What is arguably interesting in this formulation is the pervasive role of action and interaction across all three levels. In addressing the question, "why do firms have information systems," one arrives at the insight that action and interaction lie at the core of an explanation that goes beyond the simple answer, "so that they can readily do business with their customers."

With regard to the firm's need for IS, while some may view IS a mere complement to action, invoked as needed, I claim the contrary, that IS are integral to the modern firm's many day-to-day actions, which could not be carried out without them. While IS can obviously serve other roles, their primary role in firms is indisputably to serve the firm's basic actions and interactions, both at the firm level and at the coordinative subunit-level or its equivalent.

IS function in this role at the level of individual and system-level actions, where human and machine agency work together for the firm. In the case of the firms' machine systems, except where they are observed in interaction with their human users, they carry out their work largely behind the scene, in interaction with each other and with the machine systems of other firms. Significantly, in all their interactions, their application software incorporates the firm's "business rules" and their data instantiate the firm's actual practice. In this regard, the firm's machine systems tend to be more "knowledgeable" than any of their individual users. As a consequence, to the extent they are working properly, the firm's machine systems tend to be authoritative in their agency and interactions with their various human users.[9]

By way of summary, what might be termed an "IS interaction model" which brings together the basic elements of our argument is shown in Figure 2.1 below. Here the firm-level interactions between a firm and its customers and suppliers are shown by the double-head arrows. The system-level interactions are suggested by the curved lines linking the systems, as well as the spokes linking the systems to their users. The subunit-level interactions are implicit and reflected in the two internal systems with different user groups. Note that in this particular illustration, a user can access another firm's system only through his or her own firm's system, to ensure integrity of the interaction between the firms.[10]

Clearly, we have in Figure 2.1 presented only a bare-bones IS interaction model. It merely summarizes our multi-level interaction perspective. It neglects certain other interactions of importance.

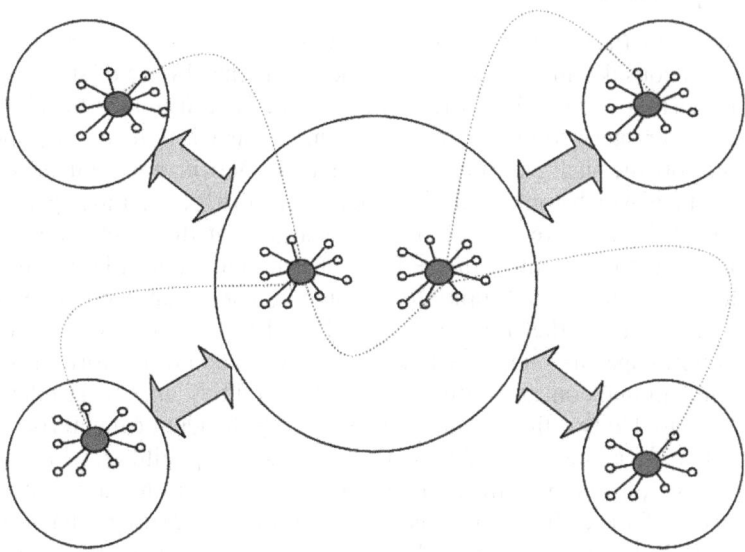

Figure 2.1 IS Interaction model.
Key: Here a firm interacts with two suppliers (to the left) and two business customers (to the right) as shown by the double-headed arrows. It coordinates its interactions with two internally linked systems, one linked also to the suppliers and one linked also to the customers. Users within and across the firms are linked to each other according to the system access they are given. System-level interactions are suggested by the curved lines linking the systems, as well as the spokes linking the systems to their users.

It excludes firm-level interactions with regulators and community organizations, for instance. It also excludes the many internal and boundary-spanning interpersonal interactions beyond the purview of the firm's machine systems. The perspective here is one in which the IS and its machine systems play the leading role in coordination of the firm's activities.

As an additional observation, while our IS interaction model is well suited to traditional firms whose customers and suppliers form distinct groups, it may also be adapted to newer enterprise forms, such as platform firms organized as multi-sided marketplaces, as with eBay. In such cases, suppliers of goods or services simply constitute a second form of customer for the platform firm, to be matched with buyers, the traditional customers. The challenge is to serve both customer groups well. With eBay, its enterprise systems, which incorporate its auction technology, facilitate the match-making.[11]

Implications

What are the implications of our argument for IS study and practice? Let's consider first the popular issue of whether IS are of strategic importance to the firm. Increasingly, some have argued, a firm's IS are characterized by commoditization, e.g., as with the use of popular off-the-shelf enterprise software, that offers no competitive advantage, which comes only from possessing distinctive and superior capabilities. However, note from our argument that such capabilities must be manifest in the form of firm and subunit-level interactions, achieved substantially through individual and system level interactions, rather than in acquired stocks of software or human "knowledge" as such. The latter must always be put to work in organizational context to display their "capabilities" and therein lays the rub. No two firms however similar with the identical enterprise software can be expected to manifest the same capabilities, which of course must be organizationally *learned.* Even with the same enterprise software, different firms will have different learning capacities and display different capabilities in their interactions with customers and suppliers, and in their own internal coordination among subunits. Even with the same enterprise software, different firms will further manifest very different individual and system-level interactions in carrying out their business processes. As a consequence, the strategic capabilities achievable with IS simply can't be reduced to the question of whether the firm's IT, such as its enterprise software, is proprietary or widely available as a commodity.[12]

For IS researchers, the implication is that organizational learning needs to be the center of attention in studies of information systems and how they support the enterprise in its operational and strategic endeavors.

As a second implication, a reframed role for human-computer interaction (HCI) studies in IS is suggested, one which goes well beyond traditional studies of human factors and designs for usability and provides a clear organizational rationale for more contemporary studies of collaborative work, e.g., that focused on the use of workflow or other groupware, or even the collective use of enterprise software. Somewhat surprisingly, notwithstanding recent work, we note that even in management oriented HCI studies, the primary research focus remains on the individual interaction, with the organization largely relegated to context. As viewed here, however, HCI provides an essential foundation for addressing larger issues, such as organizational learning and the assimilation of new technology, as just discussed.

Consider that any individual interaction in our broader framework represents but one link in what is essentially an organizational network, within which individual interactions must often be coordinated. Where coordination is overlooked, HCI studies can in effect be blinkered from organizational insights. Thus, consistent with those who have argued for the importance of coordination theory more broadly, I suggest that HCI studies that focus in particular on the coordination of business processes offer special promise for addressing the larger issues. More broadly, in attending to coordination problems, HCI studies *in situ* might be motivated and positioned in terms of the insights they can yield in enabling a firm to use IS to organizationally learn, acquire capabilities, and better do its business.[13]

As a third implication, consider that our three-level interaction framework suggests a practical approach to IS evaluation. Any particular IS may be evaluated first as to the extent to which it supports firm-level actions and interactions, secondly as to the extent to which it supports subunit-level interactions, and thirdly, in terms of its constituent individual and system-level interactions, and, in particular, the effectiveness of its HCI. One can easily imagine a three-tiered scoring system or the equivalent for periodically evaluating all of the firm's enterprise systems. The intent of such a system would be diagnostic with respect to interaction problems and issues across all three levels, such that corrective actions might be taken. For it to be effective, it would need to be built into higher-order management routines with strategic import. How this would be done would be specific to the firm.

For IS researchers, this implication suggests that multi-firm studies of IS evaluation by whatever means might be informative to the design of a three-tiered scoring system such as that suggested here.

In conclusion, in this chapter we have not only suggested an answer as to why firms have information systems, but through our analysis we have suggested too how firms may go about evaluating how well they are served by these systems. More broadly, we have seen, in particular, that an interaction perspective can shed fresh light on questions fundamental to IS study and practice. It remains to build more significantly upon this basic notion, which is merely introduced here.

Reflections

Summarizing thus far, firms are seen to have information systems to help guide their *actions* in doing business. More specifically, this guidance addresses a complex of *interactions*, at the firm level with suppliers and customers, at the sub-unit level in coordinating among functional units, and at the system level among machines and human agents engaged in organizational tasks. As will be seen in the chapters that follow, the notion of interaction is foundational to the perspective of this book. Information systems cannot be understood at any level in its absence.

Consider too that IS guidance for action in "doing business" is necessary for organizations more broadly, not only firms. Hospitals, public or private, interact with their patients, physicians, and suppliers. Libraries interact with their users and providers of documents. Charities interact with those who support them and those who are to benefit. Government agencies interact with those they serve and regulate. Even markets are typically organized around who can participate in their interactions.

The fact that doing business is commonplace to organizations of all kinds thus speaks to the centrality of information systems in the facilitation of everyday interactions among organizations and people in today's increasingly data-rich, but often information-problematic, world. Where these interactions work well, IS will often be an important reason. And where they don't, IS are likely to be at fault.

In the chapter that follows, I build on this interaction theme and venture to argue that interaction is the source of organizational information itself.

Notes

1 Alter (2002) defines an information system as a "work system that uses information technology to capture, transmit, store, retrieve, manipulate, or display information" in support of the business processes that create value for internal or external customers, a definition that itself suggests much the same answer to the question posed.

2 The notion of the value chain and its pertinence to information systems as a source of competitive advantage for the firm is described in Porter and Millar (1985). The exchange of information in business transactions gives rise to what is termed the "information value chain" which the firm may exploit to advantage, e.g., through data mining of its accumulated transaction records (Glazer, 1993).

3 Morgan (1986) presents alternative "images" of organizations, including that of the organism, and discusses the strengths and limitations of each. A problem with the organism metaphor is the assumption of functional unity. "If we look at organisms in the natural world we find they are characterized by a functional interdependence where every

element of the system under normal circumstances works for all the other elements" (p. 75). This is hardly true of organizations, whose elements may also pursue their own purposes.

4 Chandler (1977) describes the emergence in the late 19th century of the modern firm and its hierarchical form. Galbraith (1973) takes an information processing view to describe design strategies for enabling the firm to coordinate its activities within and across the hierarchy. The firm's management plays a fundamental information processing role in this regard, as do the firm's information systems. The classic paperwork vehicle employed in the early days was the multipart carbon paper form, filled out manually or by typewriter, with its parts then distributed and filed for reference among organizational subunits whose actions needed to be coordinated in carrying out the business action.

5 For background and analysis, see Malone et al. (1987), Gurbaxani and Whang (1991), and Malone (2004).

6 I acknowledge that the difference between human and machine agency is a profound one; the former is understood to bring a certain self-conscious or motivated choice to situations that has no equivalent in the latter. See, for instance, human agency in structuration theory as developed by Giddens (1984). I have myself argued that an organizational problem with machine agency in professional communications is that the personal assistant bot only feigns responsibility for its actions (Swanson, 2020).

7 What goes entirely unseen is the machine interaction within the network(s) that support the observable HCI. Here too machine agency may be at work, as with interorganizational systems that automate portions of transactions between business partners. This may be either tightly or loosely coupled through data sharing, as is increasingly the case with business ecosystems such as that which has developed around TripAdvisor as a provider of end-to-end travel services (Alaimo et al., 2020).

8 A classic coordination problem in manufacturing is between engineering and sales marketing, and the manufacturing process itself, described in Davenport (1993). Order taking and fulfillment may involve many complexities, especially where a product is made to order. Lee and Whang (2001) describe the order fulfillment process in e-commerce and how firms reliably deliver on this, especially in a holiday season.

9 The qualification, "to the extent they are working properly," is of course loaded with significance for IS practice. It may be said to constitute a perpetual challenge, one which if not met, can cripple the firm. The term "mission critical systems" invokes this crucial reality for the firm (see Davenport, 2000).

10 This is more an articulation of a design principle than a description of how many systems currently work in practice. For simplicity, the illustration also ignores the many unmediated person-to-person communications likely to be needed among system users. It further assumes that users can access only a single system. In practice, the firm is likely to be internally coordinated through dozens of interrelated, not always

integrated, systems and individual users are likely to have varied levels of access to several of them.

11 See Hasker and Sickles (2010) for a discussion of eBay's different customers and their behavior. eBay's systems must also mesh with those of its business partners, such as PayPal, which offers payment services. The emergence of business partnerships to form business ecosystems of interdependent enterprise making use of selectively shared data is one of the major developments of recent years. Here the traditional value chain is supplanted by a value added network. See, e.g., Jacobides et al. (2018) on the strategic implications of business ecosystems.

12 See, in particular, Carr (2003), who in a widely read and controversial article points to commoditization to discredit the strategic import of IT. Nelson and Winter (1982) in a classic exposition introduce the notion of organizational capabilities and discuss how they must be learned. Such capabilities are more than just the sum of the skills acquired by the organization's individual members. Significantly, individual skills must be learned so as to be effectively coordinated within broader organizational routines. It should be noted that strategic capabilities may be achieved purposefully via an articulated strategy, or "accidentally" via other pursuits, by stumbling upon a new way to improve operational effectiveness that proves to be strategic after the fact, for instance. Whether strategic advantage with IT can be sustained or not hinges on the firm's "dynamic capabilities" (its higher-order routines for purposeful ongoing adaptation), in the view of leading capability theorists (see, in particular, Teece et al., 1997; Helfat et al., 2007). Such dynamic capabilities may themselves rest upon the effective use of IT.

13 Zhang and Li (2005) provide a substantial review and assessment of HCI studies. Many management oriented HCI studies address individual IT use in the workplace, and focus on cognitive beliefs and attitudes and how they influence behavior. Surprisingly, individual learning is a relatively minor theme across more than 300 studies, addressed in only some 10% of the cases. Malone and Crowston (1994) introduce coordination theory (see also Malone, 2004). As but one illustration of promising new HCI approaches, Moran et al. (2005) describe an innovative way to coordinate business activities that seeks to mesh formal business processes with informal human collaboration. Swanson (2004) suggests that new systems must over time be organizationally assimilated through collective learning processes that combine ongoing actions with interpretation. Yamauchi and Swanson (2007) describe how workers learn on the job to use a firm's new customer relationship management (CRM) system, finding that workers do not experience HCI in isolation, but rather in close interactive relationship to each other. Successful use of the CRM hinges on these relationships.

References

Alaimo, C., Kallinikos, J. and Valderrama, E. (2020). Platforms as service ecosystems: Lessons from social media. *Journal of Information Technology*, *35*(1), 25–48.

Alter, S. (1999). A general, yet useful theory of information systems. *Communications of the AIS, 1,* 13.

Carr, N. G. (2003). IT doesn't matter. *Harvard Business Review, 81*(5), 4–11.

Chandler, A. D., Jr. (1977). *The Visible Hand.* Cambridge, MA: Belknap Press.

Davenport, T. H. (1993). *Process Innovation: Reengineering Work through Information Technology.* Boston, MA: Harvard Business School Press.

Davenport, T. H. (2000). *Mission Critical: Realizing the Promise of Enterprise Systems.* Boston, MA: Harvard Business School Press.

Galbraith, J. R. (1973). *Designing Complex Organizations.* Reading, MA: Addison-Wesley.

Giddens, A. (1984). *The Constitution of Society.* Berkeley, CA: University of California Press.

Glazer, R. (1993). Measuring the value of information: The information-intensive organization. *IBM Systems Journal, 32*(1), 99–110.

Gurbaxani, V. and Whang, S. (1991). The impact of information systems on organizations and markets. *Communications of the ACM, 34*(1), 59–73.

Hasker, K. and Sickles, R. (2010). eBay in the economic literature: Analysis of an auction marketplace. *Review of Industrial Organization, 37*(1), 3–42.

Helfat, C. E., Finkelstein, S., Mitchell, W., Peteraf, M., Singh, H., Teece, D. and Winter, S. G. (2009). *Dynamic Capabilities: Understanding Strategic Change in Organizations.* New York: John Wiley & Sons.

Jacobides, M. G., Cennamo, C. and Gawer, A. (2018). Towards a theory of ecosystems. *Strategic Management Journal, 39*(8), 2255–2276.

Lee, H. L. and Whang, S. (2001). Winning the last mile of e-commerce. *MIT Sloan Management Review, 42*(4), 54–62.

Malone, T. W. (2004). *The Future of Work.* Boston, MA: HBS Press.

Malone, T. W. and Crowston, K. (1994). The interdisciplinary study of coordination. *ACM Computing Surveys, 26*(1), 87–119.

Malone, T. W., Yates, J. and Benjamin, R. I. (1987). Electronic markets and electronic hierarchies. *Communications of the ACM, 30*(6), 484–497.

Moran, T. P., Cozzi, A. and Farrell, S. P. (2005). Unified activity management: Supporting people in e-business. *Communications of the ACM, 48*(12), 67–70.

Morgan, G. (1986). *Images of Organization.* Beverly Hills, CA: Sage Publications.

Nelson, R. R. and Winter, S. G. (1982). *An Evolutionary Theory of Economic Change.* Cambridge, MA: Harvard University Press.

Porter, M. E. and Millar, V. E. (1985). How information gives you competitive advantage. *Harvard Business Review, 63*(4), 149–160.

Swanson, E. B. (2004). How is an IT innovation assimilated? In Fitzgerald, B. and Wynn, E. (eds.), *IT Innovation for Adaptability and Competitiveness.* Dordrecht: Kluwer Academic Publishing, 267–287.

Swanson, E. B. (2020). Available to meet: Advances in professional communications. *Information Technology & People, 33*(6), 1543–1553.

Teece, D. J., Pisano, G. and Shuen, A. (1997). Dynamic capabilities and strategic management. *Strategic Management Journal*, *18*(7), 509–533.

Yamauchi, Y. and Swanson, E. B. (2010). Local assimilation of an enterprise system: Situated learning by means of familiarity pockets. *Information and Organization, 20*(3–4), 187–206.

Zhang, P. and Li, N. (2005). The intellectual development of human-computer interaction research: A critical assessment of the MIS Literature (1990–2002). *Journal of the Association for Information Systems, 6*(11), 227–292.

3 What Information Is Provided by Information Systems?

In a recent book on the subject of information, the science writer James Gleick makes an evocative observation:

> Language maps a boundless world of objects and sensations and combinations onto a finite space. ... More and more the lexicon is in the network now—preserved even as it changes: accessible and searchable. Likewise, human knowledge soaks into the network, into the cloud.[1]

In this chapter, we explore information not only in the cloud context quoted but also in what we will term the cloud of interaction.

Gaining clarity on basic concepts in a field of study such as information systems can be a challenging endeavor that sometimes resembles a cat chasing its tail. For instance, an organization's information system may be defined broadly as a computer-based system that provides information to help guide the organization's actions. While straightforward, this conception clearly begs the question of how "information" should be understood in organizational context, and in what way the computer-based system may be said to provide it. In this chapter, I venture to address the question, not so much to close on an answer, as to illuminate the ongoing chase for it.

There has recently been renewed interest in the concept of information among IS scholars.[2] Yet the concept remains elusive and marginal in the field's literature, notwithstanding a long prior history of inquiry in multiple fields of study (see especially Gleick, quoted above, for a delightful review and assessment of the information concept in the sciences). Petter and colleagues issue a call for a renewed focus:

> To ensure that the discipline has a distinguishable core, researchers often emphasize technology (or, more recently, data)

DOI: 10.4324/9781003252344-3

aspects of information systems. Yet, while the word "information" is necessary to create terms that are critical to our field,
such as "information systems" and "information technology,"
IS researchers often overlook the role of information as a component of these terms .... Information is fading from our view
of IS phenomena even though, in the era of analytics, algorithmic decision making, and "fake news," information is as appropriate to study as it has ever been or as applicable to any IS
topics under investigation.[3]

Indeed, this call is a timely one. Here I respond to it, reconsider-
ing a concept of organizational information, a particular form of
information more broadly, articulated some 40 years ago and elaborating on it in the light of newer developments. In agreement with
others, I believe that a general unified view of information is unlikely to be achievable, as underlying philosophies resist reconcilia-
tion.[4] However, researchers can position their own concepts within
the broader scheme of views. The view presented here is essentially
pragmatic, while complementary to other views.[5]

Essentially, I argue that the older concept, centered in human
communication, remains viable, but should be extended to better
incorporate the role of machines in the *actions* of organizations (the
central practical concern). The reconception is guided by several
desiderata: (i) it should be reconcilable with everyday understandings; (ii) it should comport with concepts in related fields of study;
(iii) it should incorporate organizational context; (iv) it should allow for information processing and exchange by both humans and
machines. The first two desiderata aim to keep one from straying
too far afield in the conceptualization; the second two are suggested
as necessary to contemporary computer-based systems in organizations and the future of machines, in particular.

That organizational information should be conceived as allowing for information processing by both humans and machines departs from traditional views that reserve the term "information"
for human processing and employ the term "data" for computer
processing, and where the task is to somehow convert data into information for human consumption.[6] Here I acknowledge that information processing by humans and machines is very different,
but assert that where organizations are concerned it makes sense
to consider the communicative actions taken by humans and machines in equivalent terms where they are undertaken jointly in interaction. As practice-oriented research has made clear in recent

years, both humans and machines, here understood as computers and other devices with processing and storage capability that inform their operations, serve as important actants in coordinating both the organization's work and its broader commercial and social interactions. While human agency may still be primary in establishing goals and pursuits, what the organization now routinely accomplishes follows from the purposed engagement and actions of its machines. I argue here that older concepts of organizational information need reconsideration in these terms.[7]

With all this in mind, I next revisit the original conceptualization and discussion, staying close to its text. The reconsideration follows.

Organizational Information

As background, note first that the early days of the emergent information systems field were marked by significant research addressing basic concepts, such as that of information, often published in related fields, such as management, computer science, and accounting.[8] My now old article on the "two faces of organizational information" (Swanson, 1978) appeared in a relatively new accounting journal that took a broad organizational and social view (now well established, *Accounting, Organizations, and Society* continues to be published today). Ranging widely in its discussion, the article offered a concept of organizational information that provided a new perspective.

Specifically, organizational information was seen as *purported facts given and taken, and inferences drawn and established by human participants in an organizational situation.* Such information contributes to participants' knowledge of this situation. Individual items are equivalent to natural language statements, assertions that may be taken as "true" or "false" (or "more or less true" or "more or less false") with respect to the situation. Such statements of course mean different things to different people. There is thus a certain "distribution" associated with an information item, not only in terms of its dispersion among participants but also in terms of its import to the collective knowledge held.

This particular view follows closely a classical distinction made by MacKay:

> General information theory is concerned with the problem of measuring changes in knowledge. Its key is the fact that we can

represent what we know by means of pictures, logical statements, symbolic models, or what you will. When we receive
information, it causes a change in the symbolic picture, or representation, that we could use to depict what we know.[9]

MacKay's emphasis on representation here is crucial. What is often
exchanged in organizational communication among participants,
either orally or in written form, consists of representations of purported facts or inferences. How these are accepted or not and interpreted and acted on by recipients, that is to say, how recipients
are actually themselves informed, is another matter. Human recipients will draw their own inferences in any communication exchange
(e.g., "now I know he's lying").

This view also comported with an established definition of organizational communication as "the exchange of information
between a sender and a receiver and the inference of meaning between organizational participants."[10] As noted elsewhere, the roles
of sender and receiver are further interchangeable, and, moreover,
"communication is usually reciprocal rather than unilateral, that
each participant both sends and receives within a single episode."[11]
The notion of exchange is thus suggestive of interaction as fundamental to providing information in organizational context, a point
to which we will return below.

The original conception can also be understood as elaborating
on Galbraith's seminal treatise on organization design and the
need for information, where "the greater the task uncertainty, the
greater the amount of information that must be processed among
decision makers during task execution in order to achieve a given
level of performance."[12] Here the role of computer-based systems
in organizational information processing is recognized and theo-
rized as a coordinative alternative in management, although the
information concept itself is not explicated. Much subsequent
research has built on this notion that information processing by
both humans and computer-based systems is central to coordinating the firm as a hierarchy, also without elaborating on the
concept of information itself, or how its processing by humans or
machines differs if at all.[13]

The article continues by noting that in the complex organization, it is the written record which is both source and repository
for much of the information generated and represented. However,
information is also transmitted and maintained in an organization
by word of mouth. It is not necessary for information to assume

written form before it enters into the organizational memory, as Krippendorf notes:

> Social organizations possess temporal memory by virtue of the fact that their members communicate with each other, affect each other's behavior, or participate in long chains of consequences. Information is maintained as long as it is being passed around. Naturally, processes of transmission are particularly susceptible to disturbances such as noise, additions, deletions, or super-impositions of information.[14]

That organizational information may be maintained by word of mouth only to the extent that it is repeated among participants is an interesting insight, especially since it is well known that a sequence of such repetitions may, in fact, transform the message beyond the recognition of its originator.[15] The recording of a stated fact, on the other hand, may serve to keep the original message intact. Still, there is no reason to believe that its informational nature is thereby preserved; on the contrary, it may disappear from the memories of organizational participants entirely, even to the extent of having little effect whatsoever when reintroduced through a subsequent reading. It may be then that organizational information is inherently unstable. Considerable effort and expense may be required to maintain a particular aspect of its distribution.

The original conceptualization also incorporates the phenomenon of uncertainty absorption, where inferences are drawn from a body of evidence and the inferences, instead of the evidence itself, are then communicated. Uncertainty is absorbed at such points in the sense that whether the inferences follow "reasonably" from the evidence is not examinable by the recipient of the communication, who, as a consequence, is severely limited in judging its correctness. Interpretation must be based primarily on confidence in the source and knowledge of the biases to which the source is subject, rather than on a direct examination of the evidence.[16]

This stance on information is thus multi-subject-centered and, being organizational, also sociocultural, where shared understandings may be sought, if not necessarily achieved. A virtue of the stance and conceptualization is that it is not centered in computer-based systems, but rather is human-oriented and consonant with everyday understandings of information as something more or less exchangeable (although not as "stuff") in ordinary communication.[17] It is also anchored in the research literature of the time on human and organizational communication.[18] However, from today's perspective, where computer-based systems and other digital technologies proliferate and robotics is also ascendant in the field, the concept as centered in human communication falls short in capturing organizational information processing and arguably needs to be refined and extended to incorporate machines, not only people, as organizational participants. This is illustrated most vividly in electronic commerce, where consumers make purchases online, interacting entirely with machine systems serving as the firm's transactional actants.

Having established the concept of organizational information, this article goes on to examine it primarily in terms of its import to the relationship between an organization and its environment, where information can be classified as (a) either inner- or other-directed; (b) either internally or externally based; and (c) either self- or other-referencing. It is suggested that much, if not most, organizational information is probably best regarded as "two-faced," i.e., as the product of inner- and other-directed needs taken together, with consequences for organizational self-learning and self-delusion, and for the maintenance of organizational credibility and organizational secrets. The particular role of computer-based systems in all of this is touched upon; in particular, it is noted that a computer's data base typically represents purported organizational facts, i.e., it is ostensibly "truth bearing,"[19] but the notion of information itself remains specific to human communication. In the reconsideration to follow, this is remedied.

Reconsideration

In retrospect, perhaps the most important aspect of the original view is its incorporation of interaction among organizational participants as the source of organizational information generated and maintained. In the present reconsideration, we build on this fundamentally social notion.[20] In extending the concept to include machines as participants, we suggest that interaction between humans and their machines (beyond interaction among humans) has become central to organizational information, and, more broadly, fundamental to generating and maintaining information in the contemporary built-up world.

That human-machine interaction is important in information systems has long been understood, in particular by HCI (human computer interaction) scholars. However, the original conception in management information systems (MIS) emphasized how managers or other decision-makers came to be informed by the computer-based system. Information for a system user was characterized as "data in context," for instance.[21] To the extent interaction was important, it was always in service to providing better information for the human user, whose satisfaction with the experience was central to assessing success.

The flaw in this early conception was not so much a misunderstanding of how humans could be informed by their computer-based systems, as it was the failure to grasp how computer-based systems were themselves "informed" in the interaction with users, and how, indeed, the gaining of information by the machine system, as with enterprise systems, was often at the heart of the endeavor as a whole. It thus needs to be emphasized that in any interaction between organizational participants, both parties to the interaction will be informed by it. Failure to grasp this yields a misperception of how the organization as a whole informs itself.

That machines, not just humans, may be informed in such interaction is portrayed below in a classic cartoon by Gary Glasbergen. The cartoon's humor lies in the suggestion that both parties might come to the same thought. As shown, the cloud bubble that holds this thought also captures the notion that any information gained is of the moment and rooted in the interaction itself. One is tempted to say that more broadly organizational information might be colorfully understood as gained and maintained in the cloud of interaction. Succumbing to this temptation, I chose to title this chapter accordingly, in its original conference presentation, paying homage to the cartoon.

Of course, human and machine participants are not informed in the same ways through their interactions, as is well known. In particular, human participants will gain knowledge from their awareness of the context of the communicative exchange, beyond the content of the messages they receive. Machine systems gain whatever knowledge they can from the messages themselves, interpreted through data and processing code, although some are now built to emulate humans in gaining contextual information, as with systems that monitor facial expressions of the users that interact with them.[22]

A second important aspect of the original conceptualization is the temporality of organizational information, the notion that it is generated and maintained in interaction, and that in the absence of interaction it dissipates. This notion is reflective of the view that information is not something confined to a stored representation, rather it is something that happens to someone. It contrasts with views of information as something contained intact in a written record or document, or data base, or even in a video. Instead, it regards these communicative devices as holding representations of informative potential, realizable when invoked in interaction among organizational participants, both human and machine.

The notion that the machine, not just the human, might have a temporal thought in the cloud of interaction is part of the humor in the Glasbergen cartoon and deserves comment. While I would not suggest that such a thought might fleetingly occur beyond the code and data that drives the machine's actions, I would claim that what the machine learns, beyond what it communicates, is directly comparable. For instance, it may infer that it is interacting with a novice and offer up help accordingly. It may hold this "thought" only as long as the interactive session lasts. And so for practical purposes, temporality may be built in for the machine (Figure 3.1).

A third important aspect of the original conceptualization is the attention it draws to the propagation of information in

Figure 3.1 Machines and humans informed in interaction. Reproduced with permission.

organizational context, through the "purported facts given and taken" and the "inferences drawn and established" among participants. Such propagation is regarded as important to organizational decision-making and action. In the early days of centralized computing, the machine played little role in such propagation, beyond spitting out stacks of printed reports for distribution and making a data base accessible to multiple users. Today, however, the machine has become a central player, extending its reach beyond internal use to guide interactions with a firm's suppliers and customers, for instance. It has achieved further importance as social media have come to dominate the larger scene, becoming a vehicle for organizing itself, in the interest of politics, for instance, beyond simply serving a communicative function among those already organized. The notion of organizational information has in effect been vastly extended.

In summary, the reconsideration of the original conception leads to the suggestion that machines, not just humans, be included as participants in what may be portrayed on the whole as an *open interaction network* in which organizational information is generated, maintained, and propagated to guide actions. While human-machine interactions constitute a principle portion of this network and are of central interest to information system design, interactions between machines are also an important feature, as are interactions between humans that extend information bounds beyond the domain of the machine. The network is open in the sense that it always subject to extension beyond any attempts to bound it, and, moreover, achieving scope and scale in sharing and propagating certain facts and inferences may be a collective goal.

Discussion

What emerges from the above reconsideration is a rather new view of organizational information and the systems built to provide for it. In a nutshell, this view suggests first that the purpose of an organizational information system is to support the organization in its actions, by informing and coordinating an open interaction network of people and machines. Second, it suggests that both people and machines are governed in this network through their respective agencies and knowledge, which are natural to people but artefactual to machines. Third, it suggests that organizational information generated and propagated by means of an open interaction network serves to advance organizational knowledge in the aggregate.

The new view combines representation with adaptation in one recent information taxonomy.[23] As already discussed, representation through language in which purported facts and inferences are expressed forms a foundation for interaction among human and machine participants. The meaning extracted from such interaction constitutes the in-forming process that shapes participant understandings.[24] But as suggested here, it is not just in the meanings that information has its force but also in the cloud of interaction itself, driven by organizational pragmatics, where maintenance and adaptation of the open interaction network itself is necessary. Thus, for instance, the informational import of say a firm's ERP system is to be found in the work that gets done in the multitude of interactions, more than in the semantics of the software code and database, or in the interpretations of human participants.

We note that bringing people and technology together in interaction networks is a focus of much current research that complements our own view, although it does not usually speak to the information concept.[25]

It remains to address the implications of the reconsidered view for information systems research. In doing this, we follow the insights above that organizational information arises in interaction among humans and their machines, that it is substantially temporal, and that it is generated, maintained, and propagated to guide organizational actions.

A first implication is that interaction becomes a basic focus of IS research at multiple levels beyond traditional HCI studies and that in this research it also becomes a basic unit of analysis. The smallest (micro level) unit would be the dyadic interaction between two participants, while the largest (macro level) would be a "full" interaction network. Whether micro, macro, or mezzo (in-between), the time frame of analysis of the interaction might be brief or extended, depending on the research question and study design.

A related implication is that in studying any interaction, how all parties to the interaction are informed by it is a central concern. Traditional "one-way" studies, for instance, certain usability studies that examine how a user is informed by a machine, but not the converse, are seen as having less potential for insight than those studies where learning by both parties is conjectured to be central to useful interaction. Similarly, studies of ongoing information system use that focus only on learning by the human user are less likely to yield insights than those that also address how the machine system learns (or not) from the interaction.

The second principal implication is that the temporality of organizational information becomes a potentially major focus of IS study. Following the discussion above, it is understood that information among participants can dissipate unless maintained in interaction. An important fact or inference may need reinforcement to retain salience for action. At the same time, much that is unimportant can be allowed to drop away and be forgotten. What is largely unexplored here is the role of the machine in all of this. How does the machine in interaction support the enterprise in maintaining its information, given its temporality? Inspiration here can be found in examining the design and maintenance of Wikipedia as an organized effort, where entry contributors and editors employ Talk pages to discuss and argue ideas, resolve disagreements, and shape the narrative. Importantly, Wikipedia maintains a complete developmental history of each entry and changes to it, as well as the Talk pages pertaining to it.

The third principal implication of the reconsidered view is that information propagation also becomes a potential major focus of IS study. With the advent of social media, such propagation has of course assumed high organizational importance, as reflected in a firm's use of these media not only to place ads and attract new customers but also to engage existing customers interactively, e.g., in product design, so as to build loyalty to its brands. Here, the need to build and expand an open interaction network through information propagation is a lesson already learned in practice, and research studies are well focused on how best to accomplish this. What is less well understood is how information propagation can best be contested, as illustrated not only by the problematic spread of "fake news" in the public arena but also by false or unsubstantiated claims made by customers or others about a firm's products or services, which come to wide attention.

A related issue pertains to the internal propagation of misinformation within an organization, where a particular inference about the enterprise and its situation is wrong and does not square with the facts, but is perpetuated because some or even many want to believe it. This is illustrated, for instance, where management promotes beliefs that the firm's products are safe when they are not, or that its industrial pollutants are unproblematic when they are not, or that its hiring practices are non-discriminatory when they are not. How information systems might be designed to help contest the spread of misinformation internally, not only externally among customers and others, would seem to be a worthy avenue for future research.

Application

Given these broad implications, where might we look to further explore and apply our perspective? In what organizational contexts might the concept of organizational information in an open interaction network have particular relevance for researchers and practitioners? On the face of it, we suggest that it is where *organizational reasoning* (in terms of purported facts given and taken, and inferences drawn and established) is both problematic and central to organizational action as exemplified in certain organizational routines.[26] As is well known, where organizational reasoning is relatively straightforward, even if complex, routines may be substantially automated, as with transaction processing in e-commerce. But elsewhere in a practice, reasoning may be more problematic, even if relatively routine, as in recruiting and hiring new junior employees (Goldman Sachs reportedly receives a quarter million applications annually).[27] It is here, in such problematic settings, that AI is currently making controversial inroads and where we might focus our research attention.

Consider, for illustrative example, the practice of college admissions at both the undergraduate and graduate levels. In the United States, at the undergraduate level, a "holistic approach" is increasingly undertaken by leading schools to evaluate large numbers of applications through routines featuring the deliberations of teams and committees, where efficiency is necessarily at a premium.[28] The University of Rochester, for instance, receives some 20,000 applications, and gives each a two-person team reading, followed by a committee meeting to resolve disagreements. By its nature, the holistic approach excludes admissions decisions made simply by algorithmic processing of a predefined data set. Decisions are inherently problematic in their reasoning and likely to feature information processing in an open interaction network.

Still, admissions decisions are also made in the context of organizational routines and systems that support them. They are substantially structured and guided by machine. Most schools manage their admissions using an enterprise system such as Slate, provided by Technolutions, which constitutes a "comprehensive CRM" for the review and management of admissions interactions, communications, applications, test scores, relationships, and associated materials. As promoted on its web site, Slate offers some 172 functional features for incorporation into a full range of admissions routines.[29] My own school is a user of Slate for its MBA admissions and for illustrative purposes I briefly examined this usage from the viewpoint of the open interaction network. For the Fall of 2019, UCLA Anderson enrolled 360 new full-time MBAs admitted from 2,817 applications received over the previous year. I focused on the organizational routines for recruiting these applicants and deciding upon and making offers of admissions.

The admissions process as a whole engages two principal populations in information-seeking interactions, applicants (both prospective and actual) and school representatives (admissions officers and staff supported by selected faculty, current students, and alums). From the school's perspective, the information sought pertains to the preparation, qualities, and potential of the prospect as conveyed primarily in a formal application and supportive documents, and in an invited interview, supplemented by other communications, such as email. Slate is used first to acquire information on prospects, recruit them, and guide applications from initiation to completion. The formal application takes place entirely online. Slate then guides staff readings and evaluations of the applications, interviews where invited, and admissions recommendations, decisions and acceptance actions. Throughout, each application is placed in a workflow bin that identifies its status and next steps to be undertaken. As decisions are made, Slate is further used in managing the yield on acceptance offers to achieve the target class size. Admissions concludes only with actual Fall enrollment of the class. On the whole, Slate not only provides structure for the admissions routines but also serves as the central coordinative actant, gathering information, both facts and inferences from these facts, and making these accessible according to access privilege in the open interaction network. Organizational reasoning is highly structured by design, and network communications reflect this. For instance, invited interviews are conducted mostly by current MBA student volunteers who have access only to the candidate's CV, so as not to bias the interview with the full application. Because Slate generates a vast amount of data from the interactions (including the applicant's web page visits), it mitigates against problems of information temporality as discussed above. With respect to information propagation, it seeks to restrict it according to participants' various needs to know. The importance of respecting confidentialities is underscored through training and management of the process as a whole.

While Slate offers AI prediction functionality that might be applied to applicant data to inform admissions decisions, our school does not use it. Nor is an algorithm employed. While each application is scored by readers on several categories and an overall score is assigned, no score is computed. Extensive notes are included, including reasons for recommended decisions. While several in-house studies have probed whether decisions might be better informed by applicant data analytics, none has explored machine learning. Looking ahead, the most likely AI application might be in initial screening, as some have suggested in similar contexts.[30] However, even here, we suggest that AI may problematic for precisely the reason that admissions reasoning is itself problematic, even if imitable. In our school, with its holistic approach, a rich amount of information, not all of it recorded, is propagated in the open interaction network and machine prediction of any decision recommendation or assigned score (with the possible exception of the academic score based on grades and GMAT) would likely be unreliable. Human judgement is considered necessary and highly valued throughout the process. Too, one objective of our admissions, reflecting a foundational value, is to achieve a diverse entering class. We thus have a portfolio problem that complicates the reasoning across cases. As this also suggests, the criteria against which we should judge the success of our admissions are several and the weight given to each is likely contentious within the school. Still, as Agrawal and colleagues suggest in a hypothetical example, the future for machine learning in admissions has some promise for improving upon the human attention given to different parts of the business process and we suspect some schools are experimenting with it.[31]

In summary, our illustrative case suggests new avenues for research on information in open interaction networks across a wide range of settings. Our primary suggestion is to explore organizational reasoning and its problematics more deeply where it involves machine actants. Systems such as Slate are everywhere now and provide a wealth of data to support such research using both quantitative and qualitative methods. Might the tracing, surfacing, and documenting of reasoning in certain organizational settings, such as college admissions, help to explain or even guide it as it takes place among participants? Might analysis yield insights that aid in improving upon the routines, as well as the reasoning, in future performances? Might it also enable the organization to examine and reflect upon the values and premises inherent to its practices and manifested in its routines? What forms should these analyses take in practice? More broadly, what symbiosis should organizations seek to achieve when they incorporate machine systems such as Slate in open interaction networks?

Conclusion

In conclusion, here I have sought to respond to the recent call for IS researchers to pay renewed attention to the basic concept of information. I have done so in a focused way, reexamining an older concept and suggesting that it retains viability when extended to incorporate machines, not only humans, as information processing participants in an open interaction network. The reformulated concept is at heart organizational, rather than psychological or computational. Fundamentally, it addresses organizational reasoning and in this way also responds to a recent call for research that brings a symbolic action perspective to organizational action.[32] What I have not done here is bring new insights into human information processing, or probed into how machine information processing is fundamentally different. These basic and important issues I have left to others. Rather, I have argued that where information is understood as purported facts given and taken, and inferences drawn and established among organizational participants, humans and machines can be brought together to explicate what takes place in their interactions. More broadly, the stance on information taken is both pragmatic and sociocultural in the consequentialist framework of Boell, where information is seen to exist only within a particular sociocultural background and cannot be separated from it. What is considered as "significant, technologically possible, or the things for which one has labels, categories and words, and thus can connote information"[33] will change from setting to setting. Here the knowledge that "soaks into the network, into the cloud" in Gleick's words above, soaks into the open network of interaction.

Reflections

Having argued here that interaction is the source of organizational information itself, in the next chapter I explore the topic of HCI and certain broader forms of interaction it commonly supports. While all of these interactions entail the exchange of information, many have other primary purposes, meeting other needs, such as social ones. We will see that this helps explain why everyone, in the everyday, is now an information system user. Too, we will see that who learns what from these interactions is a fundamental issue to be addressed in the information systems that facilitate them.

46 *What Information Is Provided by IS?*

Notes

1 Gleick (2011, p. 419).
2 Important contributions to the discussion include those by McKinney and Yoos (2010, 2019), Mingers and Standing (2017), and Boell (2017).
3 Petter et al. (2018, p. 10).
4 These others include Boell (2017) and Emamjome et al. (2018).
5 See, for recent example, Demetis and Lee (2019). Our own view is essentially pragmatic in the framework of Emamjome et al. (2018). See especially Goldkuhl (2004, 2012) on referential pragmatism in information systems research.
6 See Boell (2017, pp. 2–3).
7 See Latour (2005) and Nicolini (2009) on agency and actants. Brynjolfsson and McAfee (2014) provide a compelling view of the future.
8 Important contributions include those by Langefors (1980) and Stamper (1973). Contributions to the theory of information, in particular, have been international in their origins.
9 MacKay (1969, p. 42).
10 See O'Reilly and Pondy (1979, p. 121). As a subject of study, organizational communication became a division in the Academy of Management in 1974 and later absorbed IS to become the Organizational Communication and Information Systems (OCIS) division. Most recently, with some 850 members, it has renamed itself the Communication, Digital Technology, and Organization (CTO) division. It continues to provide a home for both communications and IS scholars.
11 See Weick and Browning (1986, p. 244).
12 Galbraith (1973, p. 4).
13 See, e.g., Malone and Crowston (1994).
14 Krippendorff (1975, pp. 21–22).
15 See Huber (1982).
16 March and Simon (1958, p. 165).
17 See Beynon-Davies and Wang (2019) on the problem of information sharing.
18 See, e.g., Guetzkow (1965).
19 See Mingers and Standing (2017) on data as truth-bearing.
20 See also Goguen (1997) on the social and ethical aspects of information.
21 In their important textbook, Davis and Olson (1985) define information as "data that has been processed into a form that is meaningful to the recipient and is of real or perceived value in current or prospective actions or decisions," p. 200.
22 See den Uyl and van Kuilenburg (2008) for a description of automated "face-reading."
23 See McKinney and Yoos (2010) for a taxonomy of information.
24 See Boland (1987) on "in-forming."
25 See, for example, Contractor et al. (2011).
26 See Feldman and Pentland (2003) on organizational routines.
27 See Bartleby (2018).
28 See Hoover (2017, 2018) for discussions of college-admissions issues.
29 See technolutions.com for information on Slate.

30 See, e.g., Camerer (2019), who mentions that he has sought to persuade schools to use machine learning in their admissions, but without success.
31 See Agrawal et al. (2019) on the economics of machine learning.
32 See Aakhus et al. (2014).
33 Boell (2017, p. 9).

References

Aakhus, M., Ågerfalk, P. J., Lyytinen, K. and Te'eni, D. (2014). Symbolic action research in information systems: Introduction to the special issue. *MIS Quarterly*, *38*(4), 1187–1200.

Agrawal, A., Gans, J. and Goldfarb, A. (2019). *Prediction Machines: The Simple Economics of Artificial Intelligence*. Boston, MA: Harvard Business Review Press.

Bartleby. (2018). How an algorithm may decide your career. *The Economist*, June 21, 2018.

Beynon-Davies, P. and Wang, Y. (2019). Deconstructing information sharing. *Journal of the Association for Information Systems*, *20*(4), 476–498.

Boell, S. K. (2017). Information: Fundamental positions and their implications for information systems research, education and practice. *Information and Organization*, *27*(1), 1–16.

Boland, R. J. (1987). The In-formation of information systems. In Boland, R. J. and Hirschheim, R. A. (eds.), *Critical Issues in Information Systems Research*. New York: Wiley, 363–379.

Brynjolfsson, E. and McAfee, A. (2014). *The Second Machine Age*. New York: W. W. Norton.

Camerer, C. (2019). Artificial intelligence and behavioral economics. In Agrawal, A., Gans, J., and Goldfarb, A. (eds.), *The Economics of Artificial Intelligence: An Agenda*. Chicago, IL: University of Chicago Press, 587–610.

Contractor, N., Monge, P. and Leonardi, P. M. (2011). Multidimensional networks and the dynamics of sociomateriality: Bringing technology inside the network. *International Journal of Communication*, *5*, 682–720.

Davis, G. B. and Olson, M. H. (1985). *Management Information Systems*, 2nd ed. New York: McGraw-Hill.

Demetis, D. and Lee, A. (2019). An observer-relative systems approach to information. In *Proc. of the 52nd Hawaii International Conference on System Sciences*, 6300–6309.

Emamjome, F., Gable, A., Bandara, W. and Gable, G. (2018). Re-thinking the ontology of information. *International Conference on Information Systems*, San Francisco, December 13–16, 2018.

Feldman, M. S. and Pentland, B. T. (2003). Reconceptualizing organizational routines as a source of flexibility and change. *Administrative Science Quarterly*, *48*(1), 94–118.

Galbraith, J. R. (1973). *Designing Complex Organizations.* Reading, MA: Addison-Wesley.

Gleick, J. (2011). *The Information.* New York: Pantheon.

Goguen, J. (1997). Towards a social, ethical theory of information. In Bowker, G., Star, S. L., Gasser, L., and Turner, W. (eds.), *Social Science, Technical Systems and Cooperative Work: Beyond the Great Divide,* Oxford, UK: Psychology Press, 27–56.

Goldkuhl, G. (2004). Meanings of pragmatism: Ways to conduct information systems research. In *Proc. 2nd Int. Conference on Action in Language, Organisations and Information Systems* ALOIS-2004, Linköping, 13–26.

Goldkuhl, G. (2012). Pragmatism vs interpretivism in qualitative information systems research. *European Journal of Information Systems, 21*(2), 135–146.

Guetzkow, H. (1965). Communication in organizations. In March, J. G. (ed.), *Handbook of Organizations.* Chicago, IL: Rand McNally, 534–573.

Hoover, E. (2017). Working smarter, not harder, in admissions. *Chronicle of Higher Education,* March 12, 2017.

Hoover, E. (2018). Reading an application in under 10 minutes? Way too fast, one admission dean says. *Chronicle of Higher Education,* February 1, 2018.

Huber, G. (1982). Organizational information systems: Determinants of their performance and behavior. *Management Science, 28*(2), 138–155.

Krippendorff, K. (1975). Some principles of information storage and retrieval in society. *General Systems: Yearbook of the Society for General Systems Research,* Vol. XX.

Langefors, B. (1980). Infological models and information user views. *Information Systems, 5*(1), 17–32.

Latour, B. (2005). *Reassembling the Social: An Introduction to Actor-network Theory.* Oxford, UK: Oxford University Press.

MacKay, D. M. (1969). *Information, Mechanism, and Meaning.* Cambridge, MA: MIT Press.

Malone, T. W. and Crowston, K. (1994). The Interdisciplinary study of coordination. *ACM Computing Surveys, 26*(1), 87–119.

March, J. G. and Simon, H. A. (1958). *Organizations.* New York: Wiley.

McKinney, Jr, E. H. and Yoos, C. J. (2010). Information about information: A taxonomy of views. *MIS Quarterly, 34*(2), 329–344.

McKinney, Jr, E. H. and Yoos, C. J. (2019). Information as a difference: Toward a subjective theory of information. *European Journal of Information Systems, 28*(4), 1–15.

Mingers, J. and Standing, C. (2017). What is information? Toward a theory of information as objective and veridical. *Journal of Information Technology, 33*(2), 1–20.

Nicolini, D. (2009). Zooming in and out: Studying practices by switching theoretical lenses and trailing connections. *Organization Studies, 30*(12), 1391–1418.

O'Reilly, C. and Pondy, L. (1979). Organizational communication. In Kerr, S. (ed.), *Organizational Behavior.* Columbus, OH: Grid, 119–150.

Petter, S., Carter, M., Randolph, A. and Lee, A. (2018). Desperately seeking the information in information systems research. *The DATABASE for Advances in Information Systems, 49*(3), 10–18.

Stamper, R. (1973). *Information in Business and Administrative Systems.* New York: Halstead Press (Wiley).

Swanson, E. B. (1978). The two faces of organizational information. *Accounting, Organizations and Society, 3*(3–4), 237–246.

den Uyl, M. J. and van Kuilenburg, H. (2008). The FaceReader: Online facial expression recognition. *Proc. of Measuring Behavior* 2005, 589–590.

Weick, K. E. and Browning, L. D. (1986). Argument and narration in organizational communication. *Journal of Management, 12*(2), 243–259.

4   Why Is Everyone Now An Information System User?

Taking my morning coffee and perusing the L.A. Times at the Coffee Cat not so long ago, I noticed Mary Q, a local IT consultant, working on her laptop at a nearby table. I could only imagine what Mary was up to, but had I asked, here is what she would have told me. It would not have been that different from her typical coffee-hour work routine.[1]

Upon arriving and sitting down with her latte, Mary first updated herself on several matters, going onto the Web, using local Wi-Fi access. She checked out the news at the New York Times, reading a couple of the stories, as well as the blog of a favorite columnist. She then went to the Weather Channel to get the ten-day forecast for Portland, where she had a work-related visit coming up (showers expected). She checked her G-mail, to see if there were any surprises, and finding that she needed to respond to a call for a meeting by one of her colleagues, she navigated to Doodle and made her availability known. Mulling over her upcoming business trip, she decided to book her tickets and hotel room now rather than later. She checked out alternative flights at Expedia, then went directly to United Airlines to make her choice and obtain her e-ticket. She booked a hotel room at a favorite place, going directly to the site without bothering to look for alternatives. These things accomplished, Mary was now posting news on her recent activities to her LinkedIn page.

In short, Mary was intensively engaged in what is today commonplace human computer interaction (HCI), working at her laptop. However, this barely begins to describe her interaction in the larger and more important sense in which the term is used. For while it is true that HCI by means of laptops, smartphones, tablets, and such is everywhere around us, what is also true is that this narrow-form interaction is used largely for broader-form interaction with other purposeful individuals, that is, people and organizations. It is this machine-aided broader-form interaction among

DOI: 10.4324/9781003252344-4

purposeful individuals that constitutes the important information revolution that has been with us over the last few decades, and continues with us today.

While Mary's interactions as described were indeed unremarkable on the face of it, they illustrate for us four distinct and basic forms of interaction among individuals that are the subject of interest in the present chapter. The first form is *informational*, illustrated by Mary's updating herself on the news and weather. The second is *cooperational*, illustrated by Mary's making her meeting availability known to her colleagues. The third is *transactional*, illustrated by Mary's booking of her flight and hotel room. The fourth is *social*, illustrated by Mary's updating of her LinkedIn page. What is remarkable about such varied interactions as Mary's is the almost seamless way they are now blended by IT into everyday life and work. I will term this relatively recent phenomenon the "new HCI" to contrast it from HCI as more conventionally conceived.

What I attempt to do here is take a deeper look at what is now being accomplished by the new HCI. Given that it necessarily engages multiple parties, to motivate this look I ask who learns what from it. As will be seen, I find that the learning among parties varies significantly according to the forms of interaction that motivate it. Too, third parties that increasingly facilitate the interactions themselves learn and capture value from them, driving further interaction. I argue that a new HCI perspective that focuses more explicitly on these broader-form interactions provides a promising foundation for guiding and potentially uniting future information systems research.

Forms of Interaction

As just suggested, when individual people interact with other people or organizations, these interactions may be informational, cooperational, transactional, or social. Having illustrated these forms above, here we define them and elaborate. Before doing so, we first clarify what we mean by interaction.[2]

What is important about interaction, from our perspective, is that it takes place between two or more purposeful individuals, understood as those who choose the ends they pursue, not just the means. We include not only people, but organizations as purposeful individuals.[3]

Both people and organizations can act through their machines and their human representatives, and when they interact, each party acts in response to the actions of the other. In the case of organizations, they often interact with others by means of their computer-based systems which serve as machine agents in their electronic transactions. For instance, commercial banks establish ATMs for their customers to withdraw and deposit monies. When a customer engages a bank's ATM, the principal interaction is that between the bank and the customer, while the narrower interaction between the customer and the ATM merely serves it.[4]

Because each party in an interaction is responsive to the other(s), interaction is fundamentally *informative* to participants. Each act of each participant reflects the intention that may be read by others. A certain exchange of information always takes place, even where the interaction ostensibly takes place for another purpose. Of course, the exchange is not necessarily symmetrical or fair. One party may learn much more than another through the interaction. For instance, in human interaction, one form is that of the formal interrogation, intended by one party to extract information from another, although even here, the party being interrogated may learn something about his or her interrogator through the process itself.[5]

To be more precise about an interaction, then, consider that the simplest two-party interaction with turn-taking might be modeled as P11>P21>P12>P22>P13>..., where Pij represents party i's j'th action in an interactive sequence. It is easy to imagine a two-party exchange modeled in this fashion, for instance. One sees right away that in studying interactions there may be some ambiguity in where an interaction begins and ends, although we will give this particular issue little attention here. What is more important, for our purposes, is the recognition that any interaction is fundamentally dynamic and adaptive, and that its course and outcome is largely and intentionally path-dependent. Thus, in the normal exchange, it is terminated when one or both parties decide that it should be, given where they have been and have now arrived in their exchange. Moreover, in the next exchange, where the interaction begins will likely follow from where the previous one ended, and from what the two parties have thus far learned from each other, over say a series of such interactions.[6]

Interactions as suggested by the simple model may also be conceived in studies to have certain important attributes. Among these might be the *intensity* of the interaction, as measured for instance by the length of the interactive sequence within a fixed time frame. Depending on the purpose of an interaction, its intensity maybe conjectured to be closely related to its success and to the information exchanged between the parties. Other attributes of research interest might include the interaction's *richness* (following the concept of richness employed in media richness research), *structure* (to the extent this is governed, as in a blog), and *extent* or length (the time span over which the interaction, however intense, occurs).[7]

Following our remarks above, we allow for any party to an interaction to be represented by a machine agent. This is important, for in the world in which we now live, most organizations employ the Web as a primary means of interacting with others, and this has largely given rise to the new HCI, as we have termed it. Thus, for instance, a business school may use its public website not only to provide information and guidance to prospective applicants, but to initiate and administer the formal application process itself, that is, to transact with external others.

Where organizations employ machine agents, their systems can also accumulate complete records of their machine-based interactions, which can be informative beyond the interactions of the moment, and on the Web, even within these interactions. This too is a consequence of the new HCI and indeed is now being used to extend the reach of HCI on behalf of organizations. We shall have more to say about this below, but first we elaborate on the four different forms of interaction we have already introduced above.

Informational Interaction

In informational interaction, one party seeks information from one or more others, or one party seeks to inform the actions of one or more others, or both. For example, (i) a consumer seeks product information from a manufacturer (of say appliances or automobiles). The manufacturer provides a website that makes such product information available. Or, (ii) an employee seeks retirement plan information from her employer. The employer provides periodic updates of the plan to its employees and posts these to its intranet for retrieval. Or, (iii) a service provider promotes itself via Google's AdWords, coming to the attention of a potential consumer engaged in a search, who clicks through to learn more. Or, (iv), a student unfamiliar with the concept of Web 2.0 goes online to Wikipedia to learn about it.

Informational interaction, then, revolves around the supply and demand of information. Information may be sought, that is, demanded, for all variety of reasons. It may be supplied for different reasons too. A certain exchange may take place. In particular, the information one party gains may be in return for influence sought by the other. For instance, the manufacturer which makes its product information available to consumers, even where it does not sell directly to them, hopes to influence the consumer's buying decision.[8]

In today's IT-driven world, informational interaction is most vividly illustrated by the supply of information via pages established by people and organizations on the global public Web and the corresponding demand for information of all kinds manifested by the searches of people and organizations on this same Web.

Cooperational Interaction

In cooperational interaction, two or more parties act together to accomplish a task, and information is shared and knowledge gained in the process. For example, (i) members of a design team dispersed across multiple locations use a group support system to manage and document their work. Or, (ii) a firm's supplier accesses the firm's extranet and manufacturing schedule to determine when it may need to add capacity to reliably meet demand. Or, (iii) two bank representatives, one senior, the other a trainee, cooperate in responding to a customer request, while accessing and updating the bank's enterprise system records. Or, (iv) Wikipedia volunteers from around the world interact online to maintain the encyclopedia's content.

In cooperational interaction, information is shared as a means to accomplish a collective task. Such interaction is fundamental to organizations of all kinds, where people work together, usually with some division of labor. It is important not only to formally organized traditional firms, but to loosely organized enterprises such as that illustrated by Wikipedia.[9]

Cooperational interaction lies at the heart of the area of study termed computer-supported cooperative work (CSCW), which has historically focused on creative work such as that of the design team mentioned above, but it is also more important than commonly realized to working with enterprise systems, where routines and associated data are highly structured, but where users must coordinate their efforts across multiple functions and locations.[10]

In cooperational interaction, the information shared also results over the longer term in a collective knowledge gained, that is, in organizational learning. What is learned is a capability to perform that tends to transcend attempts to represent it in storable form in so-called knowledge management systems.[11]

Transactional Interaction

In transactional interaction, two parties exchange goods, services, monies, and information pertinent to the terms of the transaction. For example, (i) a consumer buys an appliance online from a retail firm and arrangements are made for home delivery and installation. While certain of this interaction is electronic, the other is physical and specific to location. Or, (ii) an individual auctions off an item from his garage on eBay. Again, delivering the goods requires a physical process. Or, (iii) a purchasing agent for a manufacturing firm orders parts online from one of the firm's approved suppliers. Or, (iv) a trader for a bank initiates a routine trade. In this case, the entire transaction may be electronic.[12]

Much as cooperational interaction is basic to organizational operations, in terms of executing tasks, transactional interaction is basic to organizational survival and prosperity, in terms of achieving success in customer and supplier markets.

In transactional interaction, the information generated and shared typically establishes and explicates the terms of the transaction. The transactions themselves are governed by explicit or implicit contracts. The firm's accounting systems accumulate much of the basic information associated with the transactions, however, less formal information is also acquired by those individuals, such as purchasing agents, who act on behalf of the firm in executing the transactions.

Social Interaction

In social interaction, two or more parties interact with each other around their mutual interests, and information is shared in the process. For example, (i) friends on Facebook share their various "likes," including those of products or services offered by firms who have established their own Facebook presence. Or, (ii) a business school dean employs Twitter to offer her thoughts and pass along school information to her followers. Or, (iii) users of Foursquare and its social networking service are alerted to nearby places of interest on their smartphones. Or, (iv) multiple parties who do not otherwise know each other interact through their avatars in the online game of Second Life.

In much social interaction the information generated and exchanged between the parties is often incidental to the interaction itself. Participants as social beings derive direct pleasures and satisfactions from their interactions with others, as is well known. They often build out and cultivate their social networks naturally, so to speak. However, social interaction can also be exploited strategically, to advance careers, for instance, as is illustrated by professionals who use LinkedIn to network aggressively and advance themselves. Businesses have of course also developed a keen interest in exploiting social networks on the Web, to advance their products and services by digital word of mouth. They seek social recognition and acceptance of their brands, in particular.

Who Learns What

In sum, human and organizational interaction typically involves an exchange of information, mediated through the interaction itself. This exchange allows for learning by all parties. It will be helpful to examine this more closely for each of the four interaction types.

In the case of informational interaction, as we have defined it, one party seeks information from others, reflecting demand-pull, and/or others seek to provide it, reflecting supply-push. The information obtained may be more or less what was sought, and/or more or less what the provider intended that the seeker should get. Thus, what the information seeker learns from an exchange is subject to the interests and sometimes control of the provider, in particular where the provider wishes to influence the learning outcome.

For this reason, the seeker of information will also seek to ascertain and learn about the reliability of the provider. A provider may similarly seek to ascertain how subject to influence the seeker of information actually is. What is learned by both in this regard is likely to govern their future informational interactions.

Some organizations may be in the business of supporting informational interaction. On the Web, the primary example is Google, which through its search service aims to help any information seeker find a Web page pertinent to the information sought. Google thus acts as an intermediary. Its contribution to the social good has been stunning. Since its inception in the mid-1990s, Google's influence on informational interaction has been transformative, extending the global reach of seekers and providers alike.

Just as the primary parties to an informational interaction are informed by it, so too of course are information intermediaries such as Google. In particular, where the interaction takes place over the Web, because it involves information provided via stored forms, and because the interactions themselves can similarly be recorded, the intermediary can accumulate its own store of information and make it useful to others.

Cooperational interaction, like informational interaction, also provides for participant learning. However, it further provides for collective learning beyond individual learning. As mentioned above, this is in part because a substantial amount of the learning is tacit.

Cooperational interaction is also fundamental to learning by novices, who apprentice themselves and learn by doing in cooperation with experts. Because organizations are always renewing themselves with new hires as they expand or replace those departed, cooperational interaction is basic to the maintenance of organizational knowledge of many kinds. While firms have sought to capture this knowledge with knowledge management systems, because much of the knowledge is tacit, this has proved to be more difficult than envisaged.

In transactional interaction, the learning that takes place may often be more easily explicated than with cooperational interaction, reflecting a more "arms-length" relationship between participants. Each party to the transaction may come to know rather precisely what information must be exchanged and how in order to complete the transaction. Still, while each party may also learn much about their trading partners through these exchanges, there is likely to be much more that remains masked from view.

Certain information in a transactional interaction is likely to be shared. Where the transactional interaction is commercial, the information exchanged may be explicit with regard to the terms of the transaction. The sharing of such information may be basic to enforcing the terms of the contract as necessary. Thus, when a consumer makes a purchase over the Web, he or she may be required to click a box that indicates that the terms of the contract have been read and accepted (these statements are often so elaborate that most consumers will not read them).

Where a commercial transaction takes place in a formal market, the information exchanged may further be explicit with regard to the market's actions and workings. The market acts as a third party to the transaction and seeks to make its workings transparent as may be needed to secure participation. The market is thus also a participant that can learn much from the transactional interaction that take place under its auspices, as so too, at least potentially, can government agencies that provide regulatory oversight.

Certain transactional interactions may involve information exchanged as part of the transaction. Digitally represented information is a storable and transferable good. For instance, stolen credit card numbers may be sold on the black market. The New York Times provides privileged online access to its news to its subscribers.

Even where information is not exchanged as a good, each party to a transactional interaction is informed through the other's actions. Thus, although certain information will be shared, the information acquired by one party will in general be different from that acquired by the other.

In social interaction, each participant learns from the process, but much of this is incidental to the interaction itself. Information is likely to be shared, but often asymmetrically. Information is not necessarily sought, nor is it necessarily strategically proffered. In the pure case, there is no cooperative task to be performed, nor is there a transaction to be conducted. Rather, the social exchange takes place for its own sake. Valuable social ties are built and maintained. Where the exchange takes place over the Web or through other electronic forms, however, new forms of learning are enabled.

In particular, where enterprises such as Facebook and Twitter are established to facilitate social interaction, these enterprises can learn much about the participants and their social networks. Moreover, they can selectively market the information they gather to others, such as those that deliver advertisements on the Web. Not surprisingly, much effort and money are now being spent to leverage the vast information acquired about individuals and their social networks.

Discussion

To summarize, who learns what from the new HCI varies significantly according to the forms of interaction that motivate it. In informational interaction, the learning by one or more parties follows directly from the supply and demand of information itself. In cooperational interaction, the learning is substantially organizational, derived from the parties accomplishing a collective task. In transactional interaction, the parties learn from market transactions with each other. In social interaction, learning by the parties may be incidental, driven by the social value derived from the interaction. Beyond the primary parties to these interactions, across all four forms, third parties that facilitate the interactions may themselves learn and capture value from them, much of which is itself informational and drives further interaction. And so we see that the relationship between interaction and information is a powerful one.

The late scholar Horst Rittel once remarked that "information isn't stored, it happens to someone." Less evocatively, information was more commonly defined by early IS scholars as "data in context," a rather static notion where the context was often left unarticulated. Now, many years later, Rittel's assertion may be seen to have a fresh validity. For what we can observe everywhere around us, in a world now dense with IT, is that information increasingly happens to most of us through our computer-mediated *interactions*. To elaborate on Rittel, it is interaction which we undertake such that information happens to us. In this short essay, I have attempted a closer look at this interaction phenomenon, which has broad social ramifications and has drawn the attention of scholars across fields. Here I take the perspective of information systems, in particular.[13]

Not surprisingly, a dominant view in the IS context over the years has been that of HCI. As a field of study, HCI originally focused on making computers easier to use, emphasizing the design of the interface, most typically concerned with the screen and its contents and with input devices, such as keyboard and mouse. The perspective was that of design science, anchored in psychology and engineering, with attention devoted to improving the technology for the human user. Subsequently, behavioral researchers, recognizing that new technologies often encounter problems when introduced, extended HCI studies to address individual acceptance and use in organizational settings. More recently, the social use of systems has also caught the attention of HCI researchers. A recent review finds that HCI research has "expanded beyond its roots in the cognitive processes of individual users to include social and organizational processes involved in computer usage in real environments as well as the use of computers in collaboration." The notion of socially "distributed cognition" has in particular been advanced to provide a "new foundation" for HCI research. At the same time, artificial intelligence researchers aim to make the HCI experience more like a fully human one.[14]

While these more recent developments are welcome, having extended HCI research from the inside out, so to speak, I offer in the present paper a complementary perspective, which aims to position HCI more from the outside in. For if we observe more closely what is going on around us, we see that people are interacting not so much with the technology as such, but primarily with other people and organizations. HCI is with us, but in new and powerful ways, as illustrated in the opening vignette.[15]

This new HCI and its ramifications have not yet been fully grasped. Through its ubiquity, the new HCI has also become more background than foreground in our everyday observations. Of course, where we do focus on it, we see it everywhere, in places of both work and leisure. We see it especially now in mobile forms. And while the cell phone has from the beginning been a device for enabling human interaction, in its more recent smart forms its support of interaction has been vastly extended, via the Internet and Web, with profound consequences. While interaction with the cell phone as a device remains important, it is the broader human and organizational interaction transformed through its use which suggests the need for a new HCI perspective.

In the present essay, we have sketched the beginnings of such a perspective. What then might be its promise for researchers?

First, the new interaction perspective might serve to focus theory on information systems dynamics, as distinct from statics. The conventional view of IS has long been concerned with its important artifacts, in particular, beyond its devices, its software and data. Ontologies for data modeling have accordingly attracted much foundational research attention, in particular. Much of this work has a rather static flavor, derived from the view of information as storable. An alternative, practice-oriented view, less developed, has proposed anchoring IS foundations in an organization's work systems. Through process descriptions, this view has dynamic potential, though it has arguably not yet exploited interaction as such. We suggest here that an interaction perspective is fundamental and might be complementary to both these views, offering an integrative vehicle for understanding why and how software and data come to provide informational support of individual and organizational interactions, in work, but not only in work.[16]

The new interaction perspective should also serve to focus research attention on the role of individual and organizational agency in HCI. Some current research does just this. For instance, in informational interaction, studies of the individual navigation of websites are illustrative. In the case of cooperational interaction, agent-based modeling of the performance of organizational routines is also promising. In transactional interaction, game theoretic methods are being applied to the analysis of the dynamics of markets such as eBay. In social interaction, research attention is being given to the study of "opinion leadership" and other means of generating contagion phenomena on the Web.[17]

The new interaction perspective further provides a foundation for theory that incorporates individual people and organizations at the same level of analysis, for certain purposes, rather than the present approach anchored in hierarchy and organizational behavior theory, in which organizations are usually studied at a "higher" level than are people, who are studied in organizational context. This traditional approach has been very useful to IS research, but it has at the same time been limiting, arguably undervaluing or underestimating people vis-à-vis organizations in many studies. Placing people and organizations in analytic parity allows for certain studies, such as those involving transactional interactions where the terms of the contract apply to an exchange relationship between say a person and a firm, or even those that involve social interactions that take place between people and organizations on Facebook, for instance.

Too, the new interaction perspective may motivate research that takes a more institutional view of interactions among organizations and people, recognizing in particular the rise of third-party organizations based in IS that facilitate these interactions as we have illustrated here. It may allow us to understand more deeply what motivates and constitutes an "information economy" in the sense in which the term is popularly used.

Finally, the new interaction perspective also places computer-based systems in a new light, as machine agents for individuals, and organizations in particular. The new HCI is essentially one in which people interact with organizations through systems that act as machine agents. Seeking an understanding of systems as machine agents, beyond their supposedly mundane and traditional information processing roles, might also place IS research on a newer and more profound footing.

Reflections

Summarizing, we have in this chapter addressed the question of why everyone is now an information system user. We have seen that the answer lies in the recognition of different forms of HCI that address very basic human needs and that have now become features of everyday individual online life, as illustrated in the opening Coffee Cat vignette. And we have made the point that this expanded range of HCI is made possible through the organizations and information systems that provide for it.

Having thus argued here for a new perspective on HCI and the broader forms of interaction it commonly supports, in the next chapter I take a step back and examine the history of information systems and where we have now arrived with them. As will be seen, this will entail a deeper exploration of transactions, in particular, and the related interactions of organizations with people.

Notes

1 Since I first composed this essay, the Coffee Cat in Santa Barbara has closed. It was at first replaced by another coffee house where one might linger with w-fi access, as before. However, with Covid-19, this too closed. Happily, there remain other options around town.

2 While we take these four types to be basic, they are not necessarily definitive or exhaustive. Other types may be suggested, for example, recreational, political, and analytical, and no doubt still others, which we do not include as separate types here. Forms of HCI excluded from our framework include those between a single person and an isolated machine, say a primitive game station, which (re)acts on behalf of no purposeful individual.

3 See Ackoff and Emery (1972) for a treatise on purposefulness as we use the term here.

4 In our interpretation, machine agents are purposed systems (Arthur, 2009) as distinct from purposeful individuals. Their actions may commit those they represent in interactions with others, however they do not also choose the ends they themselves pursue, as do human agents. Machine agents thus represent extensions of purposeful individuals, rather than displaying full human agency as commonly understood.

5 The subject of human interaction has of course been richly researched, far beyond the minimal attention we give it here. See, for instance Blumer (1969).

6 Note that this crude modeling formulation, with turn-taking, applies best to asynchronous communications, as in texting, rather than in face-to-face communications, where simultaneity in the exchange through gestures and facial expressions may be important.

7 Daft (1986) provides an important treatment of media richness. See too Trevino et al. (1987).

8 Swanson (1992) examines the exchange of information for influence as part of a study of information accessibility.

9 Formal organization presumes cooperational interaction. Galbraith (1974) in a classic early work addressed the exchange and processing of information in organization design. More recently, Malone (2004) explores information sharing for coordinating work and in cooperative enterprise more broadly. Doan et al. (2011) discuss crowdsourcing systems on the Web.

10 See, for example, Yamauchi and Swanson (2010) for a study of a CRM system used cooperatively within and across multiple locations.

11 This capability to perform is manifest in organizational routines (Becker, 2004). We explore this "routine capability" more deeply in Chapter 6.

12 The notion that transactional interaction necessarily involves the exchange of information among parties, and that this information provides a sometimes neglected store of value, is nicely developed by Glazer (1993).

13 Davis and Olson (1985) elaborate as already noted above. Distinguishing between data and information was once popular among scholars. The quote from Rittel is from my memory of his expressing it circa 1972. Rittel is best known for introducing the concept of "wicked problems" to the policy sciences (Rittel and Weber, 1973).

14 An early classic HCI work is Card et al. (1983). The design science perspective is articulated in Carroll (1997). Zhang and Li (2005) provides a review founded in behavioral research. Olson and Olson (2003) offers a review that includes the quote, taken from its abstract. Hollan et al. (2000) makes the case for HCI based on distributed cognition. Pantic et al. (2008) articulates the ambition of AI researchers to make HCI more human-centered. Grudin (2005) elaborates on the several schools of thought associated with HCI, that have largely failed to coalesce.

15 Of course, it may be argued that these broader-form interactions have always been the basis for HCI and a reasonable case can be made for this. Certainly, the new HCI has not sprung up overnight and it has deep roots in more primitive forms of computer-based interaction. Some of these roots are in information communication technology (ICT) apart from traditional information systems.

16 See, in particular, Wand and Weber (1995), which presents a view of information systems anchored in its representational power: "The deep structure of an information system comprises those properties that manifest the meaning of the real-world system the information system is intended to model" (from the abstract). In principle, dynamic as well as static properties of real-world systems might be modeled. Alter (1999) argues for a theory of information systems based in their support of work systems.

17 Website navigation produces substantial "path data" for subsequent marketing analysis, as described by Hui et al. (2009). Its dynamics are also of intense interest to delivering advertisements in realtime. With regard to cooperational interaction, agent-based modeling of routine performance is suggested by Pentland et al. (2010). In the transactional sphere, Zeithammer (2006) uses game theory to examine the dynamics of bidding on eBay. With respect to online social interaction, Szabo and Huberman (2010) examine the dynamics by which the content of Digg and YouTube achieves popularity.

References

Ackoff, R. L. and Emery, F. E. (1972). *On Purposeful Systems.* London: Tavistock.

Alter, S. (2002). *Information Systems*, 4th ed. Upper Saddle River, NJ: Prentice-Hall.

Arthur, W. B. (2009). *The Nature of Technology.* New York: Free Press.

Becker, M. C. (2004). Organizational routines: A review of the literature. *Industrial and Corporate Change, 13*(4), 643–678.

Blumer, H. (1969). *Symbolic Interactionism: Perspective and Method.* Berkeley: University of California Press.

Card, S. K., Moran, T. P. and Newell, A. (1983). *The Psychology of Human-Computer Interaction.* Hillsdale, NJ: Erlbaum.

Carroll, J. M. (1997). Human-computer interaction: Psychology as a science of design. *Annual Review of Psychology, 48,* 61–83.

Daft, R. L. (1986). Organizational information requirements, media richness and structural design. *Management Science, 32*(5), 554–571.

Davis, G. B. and Olson, M. H. (1985). *Management Information Systems,* 2nd ed. New York: McGraw-Hill.

Doan, A., Ramakrishnan, R. and Halevy, A. (2011). Crowdsourcing systems of the World Wide Web. *Communications of the ACM, 54*(6), 86–96.

Galbraith, J. R. (1974). Organization design: An information processing view. *Interfaces,* 4(3), 28–36.

Glazer, R. (1993). Measuring the value of information: The information-intensive organization. *IBM Systems Journal, 32*(1), 99–110.

Grudin, J. (2005). Three Faces of Human-computer Interaction. *IEEE Annals of the History of Computing,* October–December, 1–18.

Hollan, J., Hutchins, E. and Kirsh, D. (2000). Distributed cognition: Toward a new foundation for human-computer interaction research. *ACM Transactions on Computer-Human Interaction, 7*(2), 174–196.

Hui, S. K., Fader, P. S. and Bradlow, E. T. (2009). Path data in marketing: An integrative framework and prospectus for model building. *Marketing Science, 28*(2), 320–335.

Malone, T. W. (2004). *The Future of Work.* Boston, MA: HBS Press.

Olson, G. M. and Olson, J. S. (2003). Human-computer interaction: Psychological aspects of the human use of computing. *Annual Review of Psychology, 54,* 491–516.

Pantic, M., Nijholt, A., Pentland, A. and Huanag, T. S. (2008). Human-centred intelligent human-computer interaction (HCI²): How far are we from attaining it? *Int. Journal of Autonomous and Adaptive Communication Systems, 1*(2), 168–187.

Pentland, B. T., Feldman, M. S., Becker, M. C. and Liu, P. (2010). The temporal foundations of organizational routines. *Helsinki Conference on the Micro-level Origins of Organizational Routines and Capabilities,* 19–20.

Rittel, H. W. J. and Weber, M. M. (1973). Dilemmas in a general theory of planning. *Policy Sciences, 4*(2), 155–169.

Swanson, E. B. (1992). Information accessibility reconsidered. *Accounting, Management and Information Technologies, 2*(3), 183–196.

Szabo, G. and Huberman, B. A. (2010). Predicting the popularity of online content. *Communications of the ACM, 53*(8), 80–88.

Trevino, L. K., Lengel, R. H. and Daft, R. L. (1987). Media symbolism, media richness and media choice in organizations. *Communications Research, 14*(5), 553–574.

Wand, Y. and Weber, R. (1995) On the deep structure of information systems. *Information Systems Journal, 5*(3), 203–223.

Yamauchi, Y. and Swanson, E. B. (2010). Local assimilation of an enterprise system: Situated learning by means of familiarity pockets. *Information and Organization, 20*(3–4), 187–206.

Zeithammer, R. (2006). Forward-looking bidding in online auctions. *Journal of Marketing Research, 43*(3), 462–476.

Zhang, P. and Li, N. (2005). The intellectual development of human-computer interaction research: A critical assessment of the MIS Literature (1990–2002). *Journal of the Association for Information Systems, 6*(11), 227–292.

5 How Did Information Systems Come to Rule the World?

Today there is much excitement around advances in artificial intelligence and robotics and how these technologies are likely to change our lives for better or worse in the years ahead. In one provocative reflection on the future, a recent New Yorker cover (October 23, 2017) portrays a busy sidewalk of robot pedestrians, one of which tosses small change into a cup held by a poor solitary human, maybe someone like us, seeking a handout. Is this where we are headed?

While the robotics revolution may thus be underway, another related and enormously important revolution has already rather quietly taken place over the last seven decades or so, a rather mundane transition that in its thoroughness and consequences has gone largely unacknowledged, even by the many who have been engaged in it over the years. I am speaking of the revolution in modern *information systems*, computer-based systems that provide information to organizations to help guide their actions.[1]

Today, it is fair to say that information systems (IS) have come to rule the world, a claim that may discomfort even some of my academic colleagues in the field, not to mention those elsewhere who have given little or no attention to what has transpired. Perhaps understandably, as always, most practitioners and academics have their eyes on the current big thing in the field, rather than on the cumulative accomplishment. Here I focus on the latter.

To be clear at the outset, when I say that IS have come to "rule" the world, I do not mean that IS rule as do kings or other authorities who may dictate to others according to personal inclination or whim. Rather, I mean that IS rule more literally, by the rules they actually embody, such that they dictate how much of everyday life, as it relates to individuals and organizations, takes place around the globe. They rule mostly without drama as infrastructure that comes to our attention only when something goes awry (which of course it does, such that glitches become something we are not surprised by).[2]

DOI: 10.4324/9781003252344-5

The everyday life I refer to involves our individual interactions with organizations and with each other, and, in particular, the *transactions* we necessarily engage in as we go about our personal and working lives. Most commonly, we transact as individuals whenever we do business in a marketplace, in whatever contemporary form, online or in a shop or mall, much as town and country folk have done since medieval times and before. But we transact too in other ways, for example, when we apply for and accept employment, or admission to a university, or when as students we register for classes, or when we commute by public bus, or attend the local theater. Many of our everyday transactions entail the payment of fees for services. A few may call for a formal agreed-upon contract, as when we engage others to renovate our house. Broadly, *both persons and organizations transact whenever they make formal, binding mutual commitments of exchange with each other.*[3] Making such commitments and carrying through on them lies at the heart of practical everyday life and of course has so for centuries. And it is in this realm that modern IS have come to serve their most basic ruling purpose, which might be termed that of *transaction facilitation*, by which we mean making it easier for mutual commitments in transacting to be made and carried out.

Transactions have of course been widely studied in other fields, not only in IS. Most notably, the importance of transactions and their costs in understanding the formation of markets and hierarchical organizations has been a central concern of scholars in institutional economics, and provides a backdrop to transaction facilitation as discussed here. The principal costs of transacting involve those of search, negotiation, and enforcement, and transaction facilitation by IS addresses all of these. But the present interest is in understanding IS and transaction facilitation in everyday life, more than exploring and explaining markets and hierarchies.[4]

In short, I address in this essay how and why IS have come to a certain kind of under-recognized prominence as infrastructure among us. I argue that it is transactions that lie at the heart of IS and how they have rather quietly come to rule the world in our everyday lives. I develop this argument with a brief, stylized review of historical developments in IS, centered on transactions and their facilitation. While there are many threads to this story, I focus in particular on developments in accounting systems, enterprise systems, retail automation, and electronic commerce. The path in these developments takes us from transaction facilitation for the enterprise only to transaction facilitation for the individual person and the enterprise. Following the review, I reflect on the importance and ramifications of the IS revolution in transaction facilitation and why it seems to have gone mostly overlooked as such in the literature. I consider where we have now arrived with transaction facilitation infrastructure and the importance of our paying more attention to it in the light of current issues, such as privacy and terms and conditions in transacting on the Web. Finally, I suggest how future studies might contribute to our learning more about where we should want to be with our transaction facilitation infrastructure.[5]

How It Happened

Accounting Systems

Accounting systems were arguably the first type of modern information system to be developed, emerging from electro-mechanical punched card accounting that preceded it.[6] These systems enabled business firms to substantially automate the classical accounting associated with their business transactions, and to keep related records and enable analyses, as well as carry out payroll processing for employees. They provided for the development of management accounting, focused on the internals of the transacting organization, beyond financial accounting, in the interest of investors and regulatory oversight.[7]

Several important characteristics of transactions between parties are highlighted by these early accounting systems. First, transactions involve not only the exchange of goods, services, and monies, but also information. Consider, for example, a 1960s era manufacturing firm that purchased parts from a supplier. The buying firm would initiate the process with a purchase order that documents the purchase and informs the supplier of what is to be provided, as well as when and how it is to be received. Upon receipt of the purchase order, the supplier would prepare its own sales order, incorporating much of the same information, which served to initiate its own process of order fulfillment. The use of paper documents to carry out such information exchange prevailed for many years until the advent of interorganizational systems in the 1980s when buyers and their suppliers began to agree on standards allowing for electronic data interchange (EDI).[8]

A second important characteristic is that transactions between parties often do not involve an instantaneous swap and must accordingly be coordinated over some period of time to their conclusion. As our example illustrates, transactions between firms may require rather elaborate sequences of actions from both parties before the transaction is successfully concluded. Consider the case where the supplier is also a manufacturer and must make the purchased product to order. The progress of fulfilling the order would likely involve many steps before the finished product is shipped and the buyer is billed with an invoice that documents the money to be provided in exchange. Upon receipt of the paper invoice as well as the purchased goods and confirming by inspection that everything is in order, the buying firm would initiate its own payment order, incorporating much of the same information. Upon receipt of payment by the supplier, the transaction is at last concluded and fully reflected in the accounts of both parties.[9]

Our example highlights a third important characteristic, which is that transactions lend themselves naturally to resolution and management by modern data processing. In particular, the information originally shared through paper documents lends itself to being entered and recorded in structured data files that can be updated as progress in completing the transaction takes place. The emergence of such structured data files and the associated software for transaction processing in accounting was a foundational development in the history of IS.[10] Notably, the programming language COBOL was devised specifically to specify file structures separately from the procedures incorporating the rules for transaction processing. Many of today's legacy systems and the business rules they incorporate remain written in COBOL.[11]

And so, modern computer-based systems with structured data files and software incorporating rules for transaction processing and accounting came to replace electro-mechanical systems. Into the 1970s, punched card technology continued to be used for data input and output was typically printed, often on multi-part forms distributed to parties needing to coordinate their actions. Beyond basic accounting, other related systems also were developed, notably in manufacturing operations management, as with materials requirements planning (MRP), to enable progress in fulfilling a customer order to be tracked. The typical organization thus came to have a portfolio of systems, an "application portfolio," that informed its routine, transaction-based actions, extending the notion of accounting systems as such. However, these systems were typically not well integrated as a whole across functional units, whose actions needed to be better coordinated.[12]

Today, accounting systems are typically integrated as financial modules within a larger ensemble known as enterprise systems. Such systems came to prominence in the 1990s to address the internal coordination problem.

Enterprise Systems

Enterprise systems, as the name suggests, bring together IS modules for supporting the enterprise as a whole. While variously interpreted across different forms of enterprise, they typically support related decisions and actions in finance, operations, and human resources, and revolve around everyday transactions with suppliers and customers.[13]

The concept of enterprise resource planning (ERP), introduced in the early 1990s, specifically addressed the coordination problem. Conceived in a manufacturing context, ERP relied upon integrated databases that could be updated in real time and shared across individual software modules. While the earliest systems were mainframe-based, relational database and client-server technologies provided new platforms that enabled ERP to be adopted and diffused more widely. Particularly important was the emergence of a competitive market among application software vendors such as SAP and Oracle incorporating common ERP business rules. At the turn of the millennium, fearing the Y2K bug, many firms moved to replace their vulnerable legacy systems with purchased ERP software.[14]

The establishment of enterprise systems among firms was important in how IS would come to rule the world in three respects. First, ES extended the rules governing everyday organizational actions to all corners of the enterprise. Indeed, their purpose was to coordinate these actions within and across business processes. Worker actions would no longer be local to a functional department. Rather, they would necessarily revolve around the ES, closely guided by its rules.[15]

Second, the notion of enterprise systems was extended beyond ERP to include CRM (Customer Relationship Management) in sales and marketing, as well as Supply Chain Management (SCM) in parts sourcing and purchasing, giving renewed emphasis to the firm's business transactions beyond the coordination of internal operations. In the case of CRM, while the earliest systems were largely internally focused on sales force management, they soon incorporated notions of the customer more consistent with the CRM name.[16]

Third, the establishment of a competitive market for enterprise system software resulted in commoditization of this software, which earlier had been locally developed and idiosyncratic to each business. As the market for this software became global, so too did the everyday business rules which it incorporated. Gradually, across industrial sectors, transactions were conducted in increasingly common ways around the world.[17]

Retail Automation

While enterprise systems delivered the rules that guided routine organizational actions, including sales, and while they increasing provided for coordination of supply chains across firms, they did not themselves govern the actions of consumers in retail markets. Everyday personal transacting outside of work was only indirectly affected by IS. Developments in retail automation did, however, set the stage for the e-commerce to come.

The first important series of developments occurred in retail banking. Here, the issuance of credit cards in the late 1950s, the establishment of automated teller machines (ATMs) in the 1960s, and the issuance of debit cards in the 1970s directly affected everyday consumer life, making both cash and credit more widely available, and facilitating payments as part of retail business more broadly. The importance of payment systems in transaction facilitation in everyday life hardly needs to be underscored. Necessarily, each of these developments relied on extensions to the banks' IS.[18]

Elsewhere, point-of-sales (POS) systems would come to facilitate transactions in traditional retail outlets such as supermarkets. Barcoding technology enabled goods to be identified and tracked and checked out at purchase.[19] Credit and debit card authorization was also built in, and cash dispensing ATMs were often placed nearby. Convenience to the customer was only one side of the facilitation picture, however. POS systems were extended to become full-blown management systems for the retail enterprise, supporting marketing initiatives, and facilitating transactions with suppliers.[20]

Another important early development occurred in the travel industry, when American Airlines built its real-time SABRE system to manage customer reservations and purchases.[21] In contrast to selling goods from a physical inventory, airlines and other transportation providers offered time-specific seats from one destination to another, presenting a very different and complex array of transactional and associated management challenges, such as scheduling and capacity management.

In the early years, most airline tickets were sold at offices of the airlines themselves. The extension of SABRE to the offices of travel agents given terminals enabled customers to book their flights and obtain tickets much more conveniently than previously. The strategic value of SABRE was quickly apparent. Originally a wholly owned enterprise, Sabre Corp. was spun off as an independent business in 2000. Even before this, Sabre software was marketed to other businesses in the travel industry, including the French Railways.[22]

Related developments occurred in the entertainment industry, in particular, with Ticketron, which through various outlets sold time and destination specific seats to concert, sport, and other events. Ticketron, developed by Control Data Corporation (CDC), was from the beginning a business unto itself, and was ultimately sold as such.[23]

The importance of these early retail developments to the story of how IS eventually came to rule the world is three-fold. First, these automation initiatives highlight transactional issues associated with time and place of goods and services purchased by the individual consumer. In particular, the importance of *convenience* to the customer, whose time and place shopping options are highly situational and circumscribed, is seen to be a major influence on automation developments undertaken.[24]

Second, pending the advent of electronic commerce, the consumers often confronted these shopping and transactional issues through interactions with distributors, more than directly with providers. The IS facilitating the transactions were either extended to distributors from suppliers, or became those of the distributors themselves.[25]

Third, in some cases, these IS became so useful that they eventually became businesses unto themselves. That IS and transaction facilitation could itself be a specialized automated business in the retail marketplace would soon be underscored with the arrival and rapid full flowering of electronic commerce.[26]

Electronic Commerce

The advent of electronic commerce, the use of the Internet, the Web, and mobile apps to transact business, and its rapid growth since the turn of the millennium, constitutes the fourth major historical development of importance to the present story.[27] While e-commerce famously attracted much early excitement as to its revolutionary nature, much of this dissipated with the dot-com crash of 2001, when many new ventures failed. Notwithstanding the collapse of the speculative bubble, however, e-commerce continued to expand unabated.

Business engagement in e-commerce has been important on two fronts, business-to-business (B2B), and business-to-consumer (B2C). While the former addresses transactions in the business supply chain, the latter addresses transactions with consumers in retail markets and is consequently important to everyday individual lives beyond the workplace.[28]

B2C e-commerce had its beginnings in business websites offering promotional "brochure ware." Online sales could not reliably be made, as sites were initially unconnected to the firm's enterprise systems where inventory levels were maintained. Once this link was forged, however, with front to back end client-server technology, customers could directly make purchases, as is common today. And in this way, we see how IS with their business rules have come to guide not only everyday actions in the working world, but everyday transactions in the consumer world. The importance of this transformation to both organizations and individuals can hardly be overstated.[29]

E-commerce thus served to awaken enterprises again to their marketplace transactions, beyond the coordination of activities achieved with enterprise systems. But it did much more than that. It unleashed entrepreneurial forces and new or transformed businesses in the marketplace that aimed to directly serve consumers in new ways made possible by the Web, including providing digital goods and services. These businesses were necessarily built largely on IS foundations.[30]

Two of these businesses on the Web, Google and Amazon, are now industry giants, and it is easy to understand why from an IS perspective. Google enabled universal search of the Web for whatever purpose and became the obvious first choice of those seeking information about most anything, including consumer products and services. As a facilitator of search, Google offers a pure example of IS as an automated enterprise.[31] Amazon grew into a massive and dominant one-stop shopping site that enables consumers to locate and purchase products and services, with direct delivery from the supplier. Most importantly, from our perspective, as its many repeat customers can testify, Amazon's IS provide for the epitome in retail transaction facilitation.[32]

Elsewhere on the Web, new businesses have proliferated to help the consumer navigate a wide range of options to meet everyday needs. The names have become familiar ones. Expedia serves as an online travel agent (OTA) offering a wide range of services. TripAdvisor crowdsources reviews of hotels and restaurants internationally. Yelp provides for reviews of local businesses. OpenTable serves to book tables at restaurants. Ticketmaster and StubHub are major providers of ticket services in sports and entertainment. PayPal allows for payments. All of these businesses (and many others) operate essentially as transaction facilitators through their IS. As market intermediaries, their success hinges on providing consumer services beyond those offered by traditional businesses as the other parties to the transactions.[33]

Additionally, beyond facilitating transactions for consumers, new businesses on the Web have also changed the playing field for individual work, giving rise to what is referred to as the gig economy, offering new opportunities (as well as challenges) for the otherwise under-employed. Uber provides a leading example, its success resting on a multi-sided IS platform facilitating transactions for both customers and affiliated ride-providers. Amazon also illustrates, with its Flex program that engages volunteer drivers to deliver packages from their own vehicles.[34]

The importance of e-commerce to the story of how IS came to rule the everyday world is thus four-fold. First, e-commerce provided the individual consumer with direct global access to both providers and distributors of goods and services, greatly expanding opportunities for transacting. Second, e-commerce gave rise to new or transformed businesses that served consumers in new ways made possible by the Web, especially in providing digital goods and services. Third, e-commerce gave rise to a plethora of new market intermediaries facilitating transactions on the Web. Fourth, e-commerce also reshuffled the playing field for individual work.

And so, we see more clearly now how IS have come to rule the world in transaction facilitation, both in the organizational world, where it primarily informs the actions of the enterprise as a whole, around which worker attention is centered, and in the everyday worlds of individuals, where it informs their actions as consumers, self-employed or volunteer or home workers, and as citizens.

How It Has Been Overlooked

How is it that the IS revolution as I describe it has been largely overlooked as such? Perhaps it comes from our failure as academics and practitioners to grasp the collective importance of the particular historical developments just described. While much excitement was generated along the way of these developments, rather little of it focused on transactions and their facilitation.

Indeed, accounting systems and their transactional extensions were and still are viewed by many as mechanical and uninteresting. Among business leaders and academics, the problems of managers have always attracted the most attention. The higher the level of the manager in the hierarchy, the greater the lure to provide innovative support. And so, in the 1960s, the field of "management information systems" (MIS) was born around a concept that resonated among business school academics, in particular. While transaction processing systems (TPS) were recognized as foundational in MIS, they were conceptually positioned at the lowest level of a hierarchy incorporating operational control, management control, and, at the apex of importance, strategic planning. MIS thus wedded itself to the hierarchy of the firm. Developers pursued prestige and importance through "executive information systems" and "strategic information systems" as advanced system types. While all of this was indeed reasonable and worthwhile, TPS and underlying business transactions in IS were seen as "routine" and largely taken for granted.[35]

With enterprise systems and ERP, the original promise was better coordination of the business as a whole, rather than better engagement in transactions as such. As it turned out, however, many firms simply replaced selected home-grown accounting systems with purchased software that did much the same transaction processing. Better coordination of the enterprise continued to prove elusive. While IS scholars lavished much attention on implementation problems associated with ERP, the emphasis was more on getting it successfully installed and underway, than on how business transactions were facilitated or not, and on how this affected everyday working life.[36]

With retail automation as discussed here, it has not been a focal topic of IS research. An interesting exception is the case of SABRE, which has been closely studied. However, SABRE has been examined primarily to gain lessons in the strategic use of IT, more than to gain insights into transaction facilitation.[37]

In the case of electronic commerce, it is now something of a mundane activity for many business firms, although this does not lessen its revolutionary importance. What now preoccupies firms and those who lead and advise them on marketing is revealing, however, in particular for B2C e-commerce. Perhaps predictably, this story has come to focus on how to market products and services through e-commerce channels. In particular, with the rise of deliverable consumer valued entertainment and other digital "content" on the Web, it now focuses on advertisement in support, and to preoccupation with how the consumer navigates the Web and can be led to and within sites where arrival can be "converted" into a purchase. Much as marketing earlier addressed physical product placement within supermarkets to maximize sales, it now addresses website designs that seek to accomplish the same thing.[38]

And so, with e-commerce, individual transactions are not at all taken for granted by the business. Rather, they are rediscovered in their importance. Here, beyond the efforts of marketers to drive customers to websites and convert them into purchases, IS and other scholars do now address the importance of achieving consumer *trust* in the B2C transactional relationship. Additionally, as businesses have opened themselves up to their customers on the Web, embracing social media, researchers have identified resulting problems for marketers in maintaining and controlling their brands.[39]

Nevertheless, while businesses now focus intently on their electronic interactions with customers, what often goes missing is the broader individual everyday life side of the transactional story. Perhaps, as our historical review suggests, this has always been the case with IS. Accounting systems, enterprise systems, retail automation, and electronic commerce, and the different systems they employ and embody, have always served the enterprise housing them first of all. And research in IS has reflected this, as if it were a priority. It has focused mostly on the enterprise as a hierarchy and has on the whole lost hindsight of the broader transactional revolution that IS has brought about.[40]

This said, ongoing research in electronic markets provides something of an exception in this regard. Significant attention has been given to peer-to-peer markets and the business platforms which enable and operate these, as with Airbnb. In practice, much attention has also been given to mobile apps that arm consumers in transactions, e.g. in making price comparisons in shopping and in checking one's own credit ratings. And current developments have brought transactions themselves to the foreground again. In particular, recent advances in blockchain technology, exemplified in Bitcoin, serve to remind us of the enduring importance of transaction facilitation in the IS story that continues to play out. Among the key blockchain functions sought are transactional validity, persistence, anonymity, privacy, traceability, and immediacy.[41]

And so a lesson here is that while transaction facilitation came to largely recede from view as it was built out as IS infrastructure, it is much too important and subject to ongoing technological and social change, such as that promised by blockchain, to be left unexamined going forward, given where we now find ourselves. While it is in the nature of infrastructures to be largely unseen among those that make everyday use of them, and while change in their complex arrangements may be so gradual as to be unobservable in the moment, this does not excuse IS scholars from leaving transactional infrastructure largely unattended, as if its upkeep was unnecessary or uninteresting, or as if its future was already resolved.[42]

Where Do We Go from Here

In summary, the suggestion here is that IS have come to rule the world through their transaction facilitation. Notably, the economist Hal Varian makes a closely related point, "so mundane and obvious, it barely seems worth mentioning," in exploring how computers have come to mediate economic transactions and impact the economy. "Nowadays," Varian observes, "most economic transactions involve a computer... (which) creates a record of the transaction." Transaction facilitation follows from this basic automated action. Indeed, and as argued here, the historical roots of this observation are deep ones. More specifically, in our terms, most economic transactions are facilitated by IS. This observation is worth more than merely mentioning. The ramifications of where we have now arrived with IS are far-reaching.[43]

For instance, of particular current interest is the advent of Big Data and associated analytics, widely understood to be a transformative innovation in computing, where data assumes a new informative prominence through its cumulative patterns. While the sources of Big Data are many, we note that one of the most important is that of transaction processing and facilitation, to which the analytics are applied in support.[44]

Looking ahead, how the IS revolution in transaction facilitation has reshaped everyday life for individuals as both consumers and workers, and how it will likely continue to do so, would seem to deserve more attention than it has been given. In concluding, I raise a few questions and offer a few thoughts on this. If IS now rule the world, how do they make it better for us or not? Who benefits from our now pervasive systems and who bears the costs? Where do we go from here and what research might be suggested to help us get to where we should want to be?[45]

First, we should note that perhaps because they serve first of all as agential extensions of enterprises, IS are deemed successful or not largely from the view of the organization that houses them. And the rules IS embody are those by which the enterprise wishes to play in transactions with other enterprises and individuals, given prevailing circumstances. These rules may be set, for instance, to differentiate between favored and disfavored suppliers, or between profitable and unprofitable customers. Indeed, SCM and CRM software, mentioned above, may be employed to do just this.[46]

Of course, the rules that IS embody are not chosen by the enterprise without some regard for their wider consequences. In B2B commerce, there may be organized efforts to arrive at industry standards for transaction facilitation, employing web-based XML to specify machine-readable content, for instance. In B2C commerce, transaction facilitation may be governed through regulatory requirements set in the interest of the consumer, and through the establishment of common legal and other protections for both the business and the consumer. Even in the absence of such regulation, businesses today must attend to the broader social acceptance of their rules and their perceived fairness.[47]

How do the business rules embodied in IS actually come about, from a transaction facilitation perspective? How do they originate and come to be put into place? How do they become infrastructure? What communities of practice and interest play what roles in effecting ongoing adaptation and change? What are the ramifications of this change for everyday life? This would seem to be a worthwhile set of research questions to be addressed.

One straightforward suggestion is to conduct studies of transaction facilitation that take account of all parties to a transaction, not just one enterprise that hosts a facilitating IS in its own interest. Studies of the crowdsourcing of work might be particularly useful in sorting out the benefits and costs to all parties, for instance. So too might be studies of ongoing consumer access of business websites, that probed aspects of usability for transacting, beyond the ease of navigation.[48]

A second suggestion is to study government and other regulation of transaction facilitation to ascertain how this works out in terms of IS. How are pertinent regulations incorporated into business rules and monitored for compliance? How are transactions actually facilitated or not by such regulation or its absence, with what consequences? Of particular interest are regulations that govern terms of use and privacy policies in e-commerce and other activities on the Web, as IS must be designed with these in mind. How effective are these IS currently from the viewpoint of all parties to the transactions and what oversight and monitoring approaches serve everyone's interests best?[49]

A well-recognized and particularly thorny issue in transaction regulation pertains to the lack of transparency as to who has what data on individuals and whether and under what conditions it can be shared or sold to third parties. Terms and conditions reflecting associated business rules and promulgated on Web sites have arguably clouded consumer understanding more than they have informed it and are increasingly under attack as inadequate to the task. How might IS be designed to provide better transparency of these associated rules, to better inform not only consumers, but whatever regulatory initiatives are undertaken in this regard?[50]

Apart from government regulation, what roles are played by professional and other communities of practice and interest in furthering transaction facilitation by means of IS? While certain interest groups may focus on maintaining a favorable business environment without excessive regulation, others may form in the public interest. Some may exploit open technologies on the Web, as with Terms of Service; Didn't Read, which seeks to apply a fairness rating system to terms of service, and make it available to individuals as a smartphone app. What are the prospects for these and other nongovernmental interventions around transaction facilitation?[51]

A third suggestion is to study how transaction facilitation by IS impacts upon the quality of working and personal life taken together, not only separately. Of particular interest would be studies that explored work that is more transactional by its nature in the gig economy and how this meshes or not with personal life, compared to contemporary employment in a hierarchy and how it is adapted or not to similarly mesh with personal life. What role does IS and transaction facilitation play in enabling people to improve the quality of their working and personal lives taken together?[52]

Lastly, a fourth suggestion is to study the evolving power relationships in transaction facilitation, and how the informational and related power asymmetries between parties are changed through new rules, AI, and access to associated aggregated data. On the face of it, organizations with their IS would seem to be increasingly better armed and positioned than individual people in their transactional relationships. Is this really the case and largely a benign development in infrastructure with which we should be comfortable? Or do we need to give more attention to certain threats posed by these increased informational and power asymmetries?[53]

Perhaps with studies such as these and others, we might learn more about how IS have come to rule the world both for better and on occasion for worse, noticed mostly in breakdowns, in the everyday lives of us all. We might learn more about IS as transactional infrastructure and how it permeates much of our lives and requires under-recognized attention. And, too, from this we might learn how new directions in IS and transaction facilitation might work toward making the world a better place.[54]

Reflections

Summarizing thus far, in the first chapter, I briefly introduced IS as computer-based systems for providing information to organizations to help guide their actions. These systems by their nature feature intensive organizational and human-computer interactions. In the second chapter, I elaborated on the organizational interactions across three levels—firm-level, subunit level, and system-level. In the third chapter, I argued that organizational information itself—the purported facts given and taken, and inferences drawn and established in an organizational situation— arises and is conveyed in interactions as we have discussed them. In the fourth chapter, I explored the basic forms of interaction that motivate the individual's typical use of today's devices—laptops, smart phones, tablets, and such—in his or her everyday mobile life, both within the organization and without. In this chapter, I have reflected on the history of IS over some seven decades and interpreted the evolutionary path as anchored fundamentally in transaction facilitation. Again, the theme of interaction has been central to the discussion.

In the two remaining chapters, I will first take a step back to examine IS as technology more broadly, with an eye on how it is associated with human practices that are common and cross organizational boundaries. In doing this, we will see again how interaction is central to reaching a useful understanding, in this case of how human practices change with IS. In the final chapter, I address the issue of how we can make a better future for ourselves with IS.

Notes

1 This interpretation of modern IS has three important aspects. IS are understood as (1) anchored in computer-based systems, (2) developed to serve organizations, and (3) purposed to guide organizational

actions. Alternatively, IS might be seen as including older pre-computer systems, and as serving individuals, and as purposed to support individual decisions. However, such a view would obscure the modern computational and organizational nature of IS, which receives emphasis here.

2 Everyone has a favorite glitch to recall. New ones come to public attention every day. As I was writing this paper, it was reported that American Airlines was without any pilots for some 15,000 Christmas flights, because of an error in its vacation bidding system that allowed all its pilots to take the holiday off if desired. Uh-oh. (The Verge, November 29, 2017).

3 That the mutual commitments are formal and binding implies that a social order exists to oversee the transactions, for example by providing legal recourse to the parties A written record, sometimes in the form of a signed formal contract, can be important. The familiar oral expression, "My word is my bond," does not typically qualify as sufficient in much of today's world, though it remains the centuries-old motto of the London Stock Exchange.

4 See, in particular, Williamson (1979). In economics, contract theory lies at the heart of the theory of the firm (Hart, 1988). Malone et al. (1987) in a seminal article explored the ramifications of IT for electronic markets and hierarchies. Bakos (1997) discusses the emergence and role of electronic markets, including that of transaction facilitation, interpreted more narrowly than in the present chapter.

5 By a "stylized review" I mean a specialized one, by no means complete, intended to bring the historic role of transaction facilitation in IS to the surface in a novel way. Much of what is covered is familiar by itself and well covered elsewhere, so I do not belabor it. I include footnotes to elaborate and give credit, and motivate further reading.

6 Accounting for business transactions has a long history dating to the earliest days of human civilization. Its rules are well established, so it is no surprise that accounting systems were among the first IS. Norberg (1990) describes the early days of punched card accounting.

7 The first business computer was LEO (Lyons Electronic Office) put into use by the J. Lyons tea and catering company in 1951 (Caminer et al., 1997). The first computer produced in the U.S. for business applications was the UNIVAC I, the first of which was shipped to the U.S. Census Bureau in 1952, with the first business sale to General Electric in 1954. For a highly readable and concise history of computing, see Ceruzzi (2012).

8 See, for instance, Mukhopadhyay et al. (1995), which examines Chrysler's use of EDI in buying assembly parts from its suppliers.

9 The notion of transaction processing systems thus came to embrace the processing of multiple associated business events, not only customer and supplier transactions (examples include the hiring of an employee and the withdrawal of a part from inventory in manufacturing). These events too were important to accounting and management. Formally, an accounting transaction is a business event that impacts the firm's financial statements. While not all business events have immediate impacts, it can be vital to account for them.

10 Rosenberg (2013) traces the origins of the concept of data to the 17th century. Structured data were perhaps first associated with electro-mechanical punched card tabulation systems. In the case of 80-column punched cards, related fields of data were pre-defined within a card that might identify and describe an inventory item, for instance. Such structured data constituted a "unit record." A file was a collection of ordered like records, describing all items in an inventory, with cards sequenced by part number, for instance. Norberg (1990) describes several early punched card applications in business. These applications expanded greatly in the 1920s. Many of these anticipate the computer-based systems that would follow in the 1950s.

11 See Sammet (1978) for an account of the early history of COBOL. While the business rules incorporated in COBOL applications were mostly computational simple, they were often exceedingly complex in that routine processing needed to both validate data and anticipate a range of exceptional cases. Over time, the software tended to grow and elaborate. Maintaining the often resulting "spaghetti code" was a notorious challenge.

12 Jacobs and Weston (2007) provide a useful history of MRP and its successor manufacturing systems, leading to enterprise resource planning (ERP). McFarlan (1981) characterizes the portfolio approach to the management of information systems.

13 Davenport (2000) provides an early discussion of enterprise systems and their importance at the turn of the millennium.

14 Kumar and Van Hillegersberg (2000) describe the emergence and early evolution of ERP, and how it consisted mostly of transaction processing. Wang and Ramiller (2009) examine the community discourse and learning associated with ERP's adoption and diffusion. Anderson et al. (2006) find that firms turned the Y2K problem into IT investment opportunities.

15 Unsurprisingly, ES implementation was often problematic and came to be widely studied. Boudreau and Robey (2005) describe how in one case workers resisted, improvised, and reinvented in confronting an ERP and its rules.

16 One such CRM notion was the "lifetime value" of a customer. A notable aim was to obtain a unified view of the customer, for instance in banks, where different services had previously been offered independently, without regard to the customer's full set of banking needs. See Peppard (2000), Rigby and Ledingham (2004), and Wang and Swanson (2008) for more background on CRM and its adoption. Davenport and Brooks (2004) describe how SCM extends the notion of an enterprise system.

17 Pollock and Williams (2008) describe how SAP came to be the leading vendor in marketing ERP software.

18 In 1958, Bank of America became the first bank to successfully issue a credit card backed by a revolving credit system in which a card was accepted by many merchants, as opposed to merchant-issued cards accepted by only a few. ATMs as cash dispensing machines first came into use in the U.K. and Sweden in 1967 (Bátiz-Lazo, 2015). See also Bátiz-Lazo and Reid (2011) for a history. Debit cards were first issued

by banks in 1975. Visa launched its ATM network, providing "anytime, anywhere" cash around the world, in 1983. Gifford and Spector (1985) describes the early development of the CIRRUS interbank system supporting ATM and credit card transactions.

19 Morton (1994) offers a history of the development of barcoding in the form of the Universal Product Code (UPC) in the U.S. Barcoding of course became integral to transaction facilitation in support of the distribution of goods throughout the economy and remains so today. Song (2003) describes how FedEx leveraged barcoding and other IT in providing e-commerce delivery services to both businesses and consumers. Jones et al. (2005) describe how RFID (radio frequency ID) technology may eventually displace barcoding as a locational technology.

20 The predecessor to the POS was the electronic cash register. In a modern POS, the customer-facing front end is but one part of the retailing information system. Other components provide for inventory control, purchasing, and receiving and transferring of products to and from other locations. Shelf replenishment is also likely to be supported, sometimes through an interorganizational link provided to suppliers, who may initiate this themselves.

21 See Copeland and McKenney (1988) for a history of SABRE's development within the broader airlines industry. Adapted from a prior IBM term, the name has no meaning as an acronym. The all-caps were eventually dropped.

22 Hopper (1990) addresses the strategic value of SABRE. Adapting SABRE's business rules to the nationalized French railway SNCF proved to be highly problematic. See Mitev (1999).

23 See Kushner (2017). Exactly what business Ticketron was in was not always clear to its management at the time.

24 The notion of convenience and the role it plays in our everyday lives and in our transactions is probably deserving of closer study. Wu (2018) suggests that: "Convenience is the most underestimated and least understood force in the world today" and that in its appeal it also has a dark side, where it can become "all destination and no journey" for us, impoverishing our lives through simplification. Too: "For all its influence as a shaper of individual decisions, the greater power of convenience may arise from decisions made in aggregate, where it is doing so much to structure the modern economy. Particularly in tech-related industries, the battle for convenience is the battle for industry dominance." There seems little doubt that convenience is a major driver in transaction facilitation, in particular.

25 The importance of IT to distribution has a long history. See, in particular, Beniger's (1986) account of the 19th century "control revolution" that transformed mass production, distribution, and consumption in the U.S. by means of the railroads and wire communications. Interestingly, even today, Internet cables often follow railroad rights of way.

26 In B2B commerce, specialized IT business was already well established, as with payroll processing. The firm ADP had its origins in the founding of Automatic Payrolls, Inc., in 1949 (see ADP's history at www.adp.com/about-adp/history.aspx).

27 The definition of e-commerce comes from Laudon and Traver (2013).
28 B2B e-commerce has given rise to both net marketplaces and private industrial networks. The former are operated by third parties and as multi-sided platforms have faced significant challenges. See Evans and Schmalensee (2005). The latter extend a firm's enterprise systems. A notable example is Procter and Gamble's system for coordinating not only with its suppliers, but with its distributors and retailers (by means of an efficient customer response system). See Laudon and Traver (2013), Chapter 12.
29 According to one report, online U.S. retail sales are expected to be 17% of total retail sales in 2022, up from 12.7% in 2017. Structural change in the marketplace is dramatic. Among online U.S. adults, 83% bought something on Amazon in 2016. See www.businessinsider. com/e-commerce-retail-sales-2017-amazon-2017-8.
30 Netflix is a particularly good example of a business that transformed itself in delivering digital entertainment on the Web.
31 Just to elaborate briefly, the backend IS of Google search consists of a search index built by Web crawlers, while the front end IS employs search algorithms to respond to queries. See www.google.com/search/ howsearchworks. As everyday users of Google search, we transact with Google as an enterprise through its terms of service, though we might only rarely if ever reference these. Google's value proposition to us is informational, not transactional. See www.google.com/policies/terms. Google's stated mission is to "organize the world's information and make it universally accessible and useful."
32 In Chapter 4, I argued that the "new human-computer interaction (HCI)" in the age of the smart phone and the Web is marked by four purposeful forms of interaction: informational, transactional, co-operational, and social. Broadly, Google responds to informational needs, while Amazon is oriented toward transactional ones. A third industry giant, Facebook, is founded on social needs. To sustain their respective businesses, both Google and Facebook have adopted advertising revenue models, and like other media companies before them, they thus serve more than one constituency.
33 Werthner and Ricci (2004) describe how the tourism industry led the way in B2C e-commerce. Hu et al. (2008) discuss the importance of payment systems in advancing mobile e-commerce. Among the many other businesses that might be mentioned is eBay, which serves as a global marketplace for the exchange of goods, and has because of its auction technology been much studied. Another is Craigslist, which serves the search function in marketplace transacting, but not the negotiation function, as with eBay. It would be useful to develop a classification of all these and other e-businesses that might illuminate transaction facilitation as described here (I suspect something like this may already exist, though I haven't found it). See Anderson and Anderson (2002) on the rise of e-commerce intermediaries.
34 Stanford (2017) provides a historical perspective on the origins of the gig economy. Semuels (2018) reflects on her experience in driving for Amazon Flex, providing insights into the platform and how it serves Amazon first of all, arguably at the expense of its package deliverers. Greenwood et al. (2017) call for interdisciplinary research that

addresses the unfolding ramifications of the gig economy for markets, firms, and individuals.

35　See Davis and Olson (1985, p. 48). This MIS text was the most authoritative of its time, the authors stating: "Today, computerized processing of transaction data is a routine activity of large organizations. ... The current challenge in information processing is to use the capabilities of computers to support knowledge work, including managerial activities and decision making" (p. 4). See Porter and Millar (1985) on strategic IS, and Watson et al. (1991) on designs for EIS. Today, management support often takes the form of an executive "dashboard" with which to drive the business as a vehicle presumably headed somewhere. Houghton et al. (2004) provides an interesting case.

36　There are of course important exceptions, especially among scholars interested in organizational learning and issues of human agency. See, in particular, Kallinikos (2004). Wagner and Newell (2004) provide a study of how an ERP system clashed with practices in a university. Yamauchi and Swanson (2010) describe how users of a bank's CRM system struggled to assimilate it in their work. Broadly, studies of communities of practice engaged in transaction facilitation allow for a focus on everyday working life. See Lave and Wenger (1991) and Brown and Duguid (1991) on communities of practice. Keller (1999) discusses lessons learned from the ERP experience.

37　Copeland and McKenney (1988) use the SABRE case to illustrate how firms can use IT to achieve economies of scale and scope, leverage technological competence, and gain sustainable advantage by exploiting opportunities.

38　Mulhern (1997) reviews the marketing decisions and automation associated with in-store retailing prior to the advent of e-commerce. Burke (2002) reports on a survey finding that online consumers regard technology largely as a means to shopping ends, rating applications highest when they made shopping more convenient. Baye et al. (2016) describe how search engine optimization (SEO) is now used to drive traffic to retail websites.

39　Siau and Shen (2003) address the challenges of achieving trust in mobile e-commerce. Fournier and Avery (2011) discuss new challenges in branding, arguing that most businesses no longer "own" their brands as such. Rather, with the advent of user-generated content on the Web, socially engaged consumers have seized co-ownership of brands.

40　Even the extensive research on computer-supported cooperative work (CSCW) has most typically been situated in an enterprise environment, where the rewards are seen to accrue to the organization(s) as a whole. The individuals studied are typically employees doing highly specialized work. The cooperation needed is often coordinative. See, for example, Majchrzak et al. (2000).

41　See Johnson (2018). Subramanian (2018) presents a vision for decentralized blockchain-based electronic marketplaces. Such marketplaces prospectively offer an alternative to firm-controlled electronic marketplaces. Davidson et al. (2018) see blockchain as an *institutional* technology that will change the economic order as it gives rise to new forms of decentralized collaborative organization.

42 Star (1999) identifies the basic characteristics of infrastructure and how it may be studied by ethnography. IS scholars have recently given significant attention to new digital infrastructures such as those made possible by the Internet and the Web, even as traditional IS and transaction facilitation infrastructure continues to be taken mostly for granted. See, for example, Constantinides et al. (2018).

43 See Varian (2010, p. 2). Varian assesses the organizational and economic impact of computer mediated transactions in facilitating new forms of contracts, data extraction and analysis, controlled experimentation, and personalization and customization. He conjectures that "micro multinationals" will play an increasingly important role in the economy.

44 Here I simply touch on this important development. For one interesting take on Big Data and its ramifications for organizations, see Constantiou and Kallinikos (2015). Ramiller and Chiasson (2016) draw on critical theory to examine the institutional logic of Big Data.

45 Yoo (2010) has issued a related call for research focused on the role of computing in everyday life, emphasizing the importance of computers embedded in other devices, such as automobiles. He terms this new domain "experiential computing."

46 See, for example, the illuminating case of CRM's use at Royal Bank (Khirallah, 2001). Whether the bank should profit more from some of its customers than others and favor them accordingly is understood to be an issue in its use of CRM.

47 See Wigand et al. (2005) on the development of industry standards in the U.S. home mortgage industry. Jamal et al. (2003) examine the adoption of common privacy protections in e-commerce, and on how these are disclosed and assured in accounting. Social acceptance (or not) of business rules is increasingly reflected in controversies discussed on social media. For instance, whether qualifying Amazon Flex package deliverers receive their tips separately or whether these are folded into the base pay if needed has recently been questioned. One deliverer complains that there is "zero transparency about our pay" (Bhuiyan, 2019). Such cases provide grist for regulatory initiatives.

48 Such studies appear to be well underway. Crowdsourcing may allow some workers to build their careers through job crafting, for instance (Taylor and Joshi, 2016). Closer focus on the transactional aspects of crowdsourcing might be revealing of the opportunities and limitations for such career building through job crafting. How are IS designed not only to manage crowdsourcing for the enterprise, but to support job crafting and career building for the individual?

49 Laudon (1996) argues for privacy protection via individual ownership of personal information and the creation of a national market in which individuals are compensated for sharing information about themselves. Hoffman et al. (1999) address privacy and trust in shopping on the Web, arguing for a more consumer-oriented model of personal data ownership. Spiekermann et al. (2015) offer a recent assessment of prospects for personal data markets. From the present viewpoint, the focus of privacy might be on the data exchanged or not in a transaction, and how the IS facilitates this, where both parties have privacy claims.

Certain data might be private to the transaction itself and shared only by prior agreement, as with a credit card application accepted or denied.

50 A New York Times editorial reports that reading Amazon's terms and conditions out loud takes about nine hours. A recent art exhibit plays a recording of such a reading that opens each day and runs continually until the closing hour without concluding (Editorial Board, 2019). Studies suggest that people do not read terms of service, but rather have been habituated to simply click "accept" in order to proceed in their task when presented with terms over which they have no control in the moment (Berreby, 2017).

51 See https://tosdr.org/. The notion of a community of interest, that brings together individuals from multiple communities of practice around a common interest, was introduced by Fischer (2001) in a design context different from that of the Web. The volunteer and solicited work that arises to fill the cracks in our progressively automated systems has become a feature of everyday life for many (Ekbia and Nardi, 2014).

52 Autor (2015) and Spreitzer et al. (2017) examine the future of work in the post-industrialization economy. Barley and Kunda (2001) make the case for closer studies of work by organizational researchers. To address work-life balance, the notion of work might be extended to cover personal work outside of employment, much of which involves transacting that is facilitated or not by IS. The popular notion of work as paid employment balanced by "leisure" is a quaint one in the contemporary economy. How do individuals cope with current work challenges? See also Petriglieri et al. (2019), who identify emotional problems in forging personal work identities in the gig economy.

53 Much may be at stake. For a trenchant critique of where we now are with "surveillance capitalism" and the prospects for an "information civilization," see Zuboff (2015). See also Clarke (2019), which sounds a similar alarm, with the claim that "in the digital surveillance economy, genuine relationships between organizations and people are replaced by decision-making based on data that has been consolidated into digital personae" and where "The consumer is converted from a customer to a product, consumers' interests have almost no impact on the process, and they can be largely ignored" (p. 60).

54 I note here the recent formation of the Responsible Research in Business & Management (RRBM) Network, which seeks to "transform business and management research toward achieving humanity's highest aspirations for a better world." See www.rrbm.network. The present essay, centered on transaction facilitation, attempts to open one door toward that end. See also Walsham (2012) for an earlier call.

References

Anderson, M. C., Banker, R. D. and Ravindran, S. (2006). Value implications of investments in information technology. *Management Science*, 52(9), 1359–1376.

Anderson, P. and Anderson, E. (2002). The new e-commerce intermediaries. *Sloan Management Review*, *43*(4), 53–62.

Autor, D. H. (2015). Why are there still so many jobs? The history and future of workplace automation. *Journal of Economic Perspectives*, *29*(3), 3–30.

Bakos, Y. (1998). The emerging role of electronic marketplaces on the Internet. *Communications of the ACM*, *41*(8), 35–42.

Barley, S. R. and Kunda, G. (2001). Bringing work back in. *Organization Science*, *12*(1), 76–95.

Bátiz-Lazo, B. (2015). A brief history of the ATM. *The Atlantic*. March 26, 2015.

Bátiz-Lazo, B. and Reid, R. (2011). The development of cash-dispensing technology in the UK. *IEEE Annals of the History of Computing*, *33*(3), 32–45.

Baye, M. R., De los Santos, B. and Wildenbeest, M. R. (2016). Search engine optimization: what drives organic traffic to retail sites? *Journal of Economics & Management Strategy*, *25*(1), 6–31.

Beniger, J. (1986). *The Control Revolution: Technological and Economic Origins of the Information Society*. Cambridge, MA: Harvard University Press.

Berreby, D. (2017). Click to agree with what? No one reads terms of service, studies confirm. *The Guardian*. March 3, 2017.

Bhuiyan, J. (2019). Where does a tip to an Amazon driver go? In some cases, toward the driver's base pay. *Los Angeles Times*, February 7, 2019.

Boudreau, M. C. and Robey, D. (2005). Enacting integrated information technology: A human agency perspective. *Organization Science*, *16*(1), 3–18.

Brown, J. S. and Duguid, P. (1991). Organizational learning and communities-of-practice: Toward a unified view of working, learning, and innovation. *Organization Science*, *2*(1), 40–57.

Burke, R. R. (2002). Technology and the customer interface: What consumers want in the physical and virtual store. *Journal of the Academy of Marketing Science*, *30*(4), 411–432.

Caminer, D., Land, F., Aris, J. and Hermon, P. (1997). *LEO: The Incredible Story of the World's First Business Computer*. New York: McGraw-Hill Professional.

Ceruzzi, P. E. (2002). *Computing: A Concise History*. Cambridge: MIT Press.

Clarke, R. (2019). Risks inherent in the digital surveillance economy: a research agenda. *Journal of Information Technology*, *34*(1), 59–80.

Constantinides, P., Henfridsson, O. and Parker, G. G. (2018). Introduction—Platforms and infrastructures in the digital age. *Information Systems Research*, *29*(2), 381–400.

Constantiou, I. D. and Kallinikos, J. (2015). New games, new rules: Big data and the changing context of strategy. *Journal of Information Technology*, *30*(1), 44–57.

Copeland, D. G. and McKenney, J. L. (1988). Airline reservations systems: Lessons from history. *MIS Quarterly, 12*(3), 353–370.

Davenport, T. H. (2000). *Mission Critical: Realizing the Promise of Enterprise Systems.* Boston, MA: Harvard Business School Press.

Davenport, T. H. and Brooks, J. D. (2004). Enterprise systems and the supply chain. *Journal of Enterprise Information Management, 17*(1), 8–19.

Davidson, S., De Filippi, P. and Potts, J. (2018). Blockchains and the economic institutions of capitalism. *Journal of Institutional Economics, 14*(4), 639–658.

Davis, G. B. and Olson, M. H. (1985). *Management Information Systems,* 2nd ed. New York: McGraw-Hill.

Editorial Board. (2019). How Silicon Valley puts the 'con' in consent. *New York Times.* February 2, 2019.

Ekbia, H. and Nardi, B. (2014). Heteromation and its (dis) contents: The invisible division of labor between humans and machines. *First Monday, 19*(6), online.

Evans, D. S. and Schmalensee, R. (2005). *The Industrial Organization of Markets with Two-Sided Platforms.* No. w11603. National Bureau of Economic Research, Cambridge, MA.

Fischer, G. (2001). Communities of interest: Learning through the interaction of multiple knowledge systems. In *Proc. 24th IRIS Conference* (1), 1–13. Department of Information Science, University of Bergen.

Fournier, S. and Avery, J. (2011). The uninvited brand. *Business Horizons, 54*, 193–207.

Gifford, D. and Spector, A. (1985). The CIRRUS banking network. *Communications of the ACM, 28*(8), 798–807.

Greenwood, B., Burtch, G. and Carnahan, S. (2017). Unknowns of the gig economy. *Communications of the ACM, 60*(7), 27–29.

Hart, O. D. (1988). Incomplete contracts and the theory of the firm. *Journal of Law, Economics, & Organization, 4*(1), 119–139.

Hoffman, D. L., Novak, T. P. and Peralta, M. (1999). Building consumer trust online. *Communications of the ACM, 42*(4), 80–85.

Hopper, M. D. (1990). Rattling SABRE-new ways to compete on information. *Harvard Business Review, 68*(3), 118–125.

Houghton, R., El Sawy, O. A., Gray, P., Donegan, C. and Joshi, A. (2004). Vigilant information systems for managing enterprises in dynamic supply chains: Real-time dashboards at Western Digital. *MIS Quarterly Executive, 3*(1), 19–35.

Hu, X., Li, W. and Hu, Q. (2008). Are mobile payment and banking the killer apps for mobile commerce? In *Proc. of the 41st Annual Hawaii International Conference on System Sciences,* 84.

Jacobs, F. R. and Weston, F. C., Jr. (2007). Enterprise resource planning (ERP)—A brief history. *Journal of Operations Management, 25,* 357–363.

Jamal, K., Maier, M., and Sunder, S. (2003). Privacy in e-commerce: development of reporting standards, disclosure, and assurance services in an unregulated market. *Journal of Accounting Research, 41*(2), 285–309.

Johnson, S. (2018). Beyond the bitcoin bubble. *New York Times.* January 16, 2018.

Jones, M. A., Wyld, D. C. and Totten, J. W. (2005). The adoption of RFID technology in the retail supply chain. *The Coastal Business Journal, 4*(1), 29–42.

Kallinikos, J. (2004). Deconstructing information packages: Organizational and behavioural implications of ERP systems. *Information Technology & People, 17*(1), 8–30.

Keller, E. L. (1999). Lessons learned. *Manufacturing Systems, 17*(11), 44–50.

Khirallah, K. (2001). *CRM Case Study: The Analytics That Power CRM at Royal Bank [of Canada].* Needham, MA: TowerGroup.

Kumar, K. and Van Hillegersberg, J. (2000). ERP experiences and evolution. *Communications of the ACM, 43*(4), 22–22.

Kushner, S. (2017). How events tickets started to compute. Charles Babbage Institute Newsletter. At www.cbi.uman.edu/article8.html.

Lave, J. and Wenger, E. (1991). *Situated Learning. Legitimate Peripheral Participation.* Cambridge: University of Cambridge Press.

Laudon, K. C. (1996). Markets and privacy. *Communications of the ACM, 39*(9), 92–104.

Laudon, K. C. and Traver, C. G. (2013). *E-commerce*, 9th ed. Boston, MA: Pearson.

Majchrzak, A., Rice, R. E., Malhotra, A., King, N. and Ba, S. (2000). Technology adaptation: The case of a computer-supported inter-organizational virtual team. *MIS Quarterly, 24*(4), 569–600.

Malone, T. W., Yates, J. and Benjamin, R. I. (1987). Electronic markets and electronic hierarchies. *Communications of the ACM, 30*(6), 484–497.

McFarlan, F. W. (1981). Portfolio approach to information systems. *Harvard Business Review, 59*(5), 142–150.

Mitev, N. N. (1999). Electronic markets in transport: Comparing the globalization of air and rail computerized reservation systems. *Electronic Markets, 9*(4), 215–225.

Morton, A. Q. (1994). Packaging history: The emergence of the uniform product code (UPC) in the United States, 1970–75. *History and Technology, an International Journal, 11*(1), 101–111, DOI: 10.1080/07341519408581856.

Mukhopadhyay, T., Kekre, S. and Kalathur, S. (1995). Business value of information technology: a study of electronic data interchange. *MIS Quarterly, 19*(2), 137–156.

Mulhern, F. J. (1997). Retail marketing: From distribution to integration. *International Journal of Marketing Research, 14*(2), 103–124.

Norberg, A. L. (1990). High-technology calculation in the early 20th Century: Punched card machinery in business and government. *Technology and Culture, 31*(4), 753–779.

Peppard, J. (2000). Customer relationship management (CRM) in financial services. *European Management Journal, 18*(3), 312–327.

Petriglieri, G., Ashford, S. J. and Wrzesniewski, A. (2019). Agony and ecstasy in the gig economy: Cultivating holding environments for precarious and personalized work identities. *Administrative Science Quarterly*, *64*(1), 124–170.

Pollock, N. and Williams, R. (2008). *Software and Organisations: The Biography of the Enterprise-wide System or how SAP Conquered the World*. London: Routledge.

Porter, M. E. and Millar, V. E. (1985). How information gives you competitive advantage. *Harvard Business Review*, *63*(4), 149–160.

Ramiller, N. and Chiasson, M. (2016). Datafication and the production of the post-human consumer. Unpublished working paper, January 15, 2016.

Rigby, D. K. and Ledingham, D. (2004). CRM done right. *Harvard Business Review*, *82*(11), 118–29.

Rosenberg, D. (2013). Data before the fact. In Gitelman, L. (ed.), *Raw Data Is an Oxymoron*. Cambridge, MA: MIT Press., 15–40.

Sammet, J. E. (1978). The early history of COBOL. In Wexelblat, R. L. (Ed.), *History of Programming Languages I*, 199–243.

Semuels, A. (2018). I delivered packages for Amazon and it was a nightmare. *The Atlantic*, June 25, 2018.

Siau, K. and Shen, Z. (2003). Building customer trust in mobile commerce. *Communications of the ACM*, *46*(4), 91–94.

Song, H. (2003). E-services at FedEx. *Communications of the ACM*, *46*(6), 45–46.

Spiekermann, S., Acquisti, A., Böhme, R. and Hui, K. L. (2015). The challenges of personal data markets and privacy. *Electronic Markets*, *25*(2), 161–167.

Spreitzer, G. M., Cameron, L. and Garrett, L. (2017). Alternative work arrangements: Two images of the new world of work. *Annual Review of Organizational Psychology and Organizational Behavior*, *4*, 473–499.

Stanford, J. (2017). The resurgence of gig work: Historical and theoretical perspectives. *The Economic and Labour Relations Review*, *28*(3), 382–401.

Star, S. L. (1999). The ethnography of infrastructure. *American Behavioral Scientist*, *43*(3), 377–391.

Subramanian, H. (2018). Decentralized blockchain-based electronic marketplaces. *Communications of the ACM*, *61*(1), 78–84.

Taylor, J. and Joshi, K. D. (2016). Building a career: Job-crafting through IT crowdsourcing. *4th Int. Workshop on the Changing Nature of Work*, Dublin.

Varian, H. R. (2010). Computer mediated transactions. *American Economic Review*, *100*(2), 1–10.

Wagner, E. L. and Newell, S. (2004). Best for whom: The tension between 'best practice' ERP packages and diverse epistemic cultures in a university context. *Journal of Strategic Information Systems*, *14*(4), 305–328.

Walsham, G. (2012). Are we making a better world with ICTs? Reflections on a future agenda for the IS field. *Journal of Information Technology, 27*(2), 87–93.

Wang, P. and Ramiller, N. C. (2009). Community learning in information technology innovation. *MIS Quarterly 33*(4), 709–734.

Wang, P. and Swanson, E. B. (2008). Customer relationship management as advertised: Exploiting and sustaining technological momentum. *Information Technology & People, 21*(4), 323–349.

Watson, H. J., Rainer, K. and Koh, C. (1991). Executive information systems: A framework for development and a survey of current practice. *MIS Quarterly, 15*(1), 13–30.

Werthner, H. and Ricci, F. (2004). E-commerce and tourism. *Communications of the ACM, 47*(12), 101–105.

Wigand, R. T., Steinfield, C. W. and Markus, M. L. (2005). Information technology standards choices and industry structure outcomes: The case of the US home mortgage industry. *Journal of Management Information Systems, 22*(2), 165–191.

Williamson, O. E. (1979). Transaction-cost economics: The governance of contractual relations. *The Journal of Law and Economics, 22*(2), 233–261.

Wu, T. (2018). The tyranny of convenience. *New York Times.* February 16, 2018.

Yamauchi, Y. and Swanson, E. B. (2010). Local assimilation of an enterprise system: Situated learning by means of familiarity pockets. *Information and Organization, 20*(3–4), 187–206.

Yoo, Y. (2010). Computing in everyday life: A call for research on experiential computing. *MIS Quarterly, 34*(2): 213–231.

Zuboff, S. (2015). Big other: Surveillance capitalism and the prospects of an information civilization. *Journal of Information Technology, 30*(1), 75–89.

6 How Do Human Practices Change with Information Systems?

Having sketched the history of information systems and how they have come to "rule the world" today, in this chapter I attempt to establish new foundations for thinking about information systems in the context of technological change and its ramifications for change in human practices in general, and for the future of work, in particular.

The subject of technological change is an important one for information systems, as change in underlying IT has enabled their evolutionary development into many new forms over the decades since their emergence. Consider enterprise systems, for instance. In the 1990s, ERP (Enterprise Resource Planning) emerged as an important new form, as discussed in the previous chapter. Its origins dated back to the 1960s as described in Figure 6.1. ERP itself constitutes a technology in the sense that it is a means to fulfill a human purpose, as Brian Arthur has described it, a purposed system that incorporates method, process, and device.[1] But how should we understand this more deeply?

In what follows, I introduce the notion of technology as a kind of *routine capability*, arguing that routines are central to understanding what technology itself is all about. I then elaborate on origins and modes of technological change and illustrate in the case of ERP. With this in hand, I then consider the ramifications of this perspective for broader changes in work and human practices.

Technology as Routine Capability

Recent interest in pervasive digitalization has yielded important insights into our digital devices and their affordances, however this has arguably reinforced our perspective of technology as material stuff, such as most recently the Internet of Things. Just how does technology understood as stuff have organizational and social

DOI: 10.4324/9781003252344-6

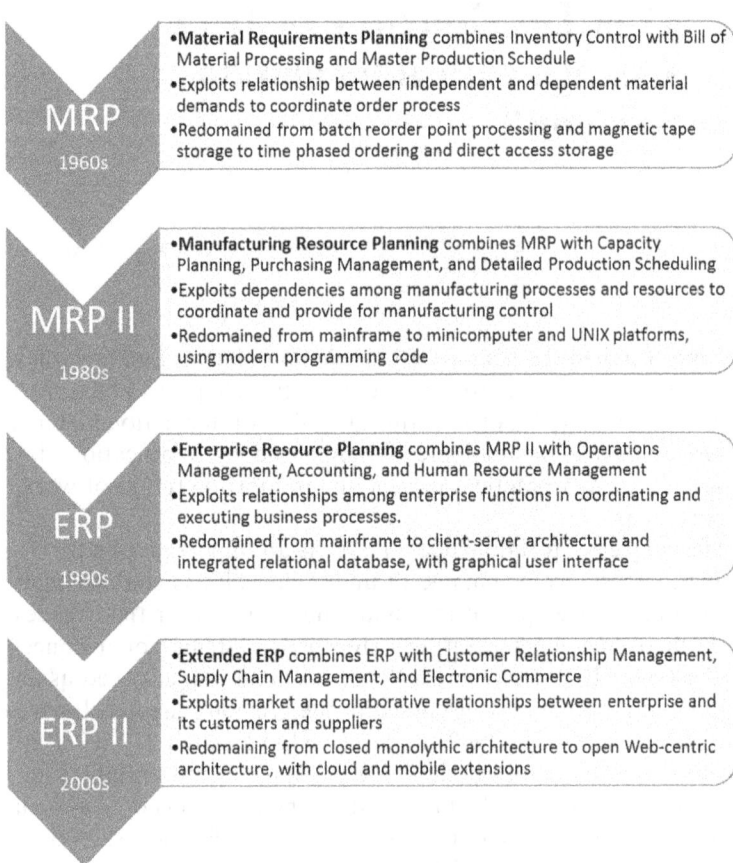

Figure 6.1 Evolution of ERP. Adapted from Swanson (2017).

"impacts," such as those on work? While significant changes in the nature of work over the years can be observed, such as the amount of time knowledge workers now spend face to screen, and readily attributed to digitalization and its devices, such as ever more powerful computers and smartphones with Internet access, how actually does all this come about and does this matter?[2]

With sociomaterial theorists and others, I argue that technology should be understood as more than devices, the common view. As seen here, technology is best understood as "routine capability" forged through actual *use* of devices. The aim of this perspective is to highlight the role of organizational routines, recurrent patterns of action among individuals in bringing technological innovation about and delivering on its promise to people and their organizations. The importance of routines to organizations has long been understood and has been much researched. The present essay asks whether routines might in this broader innovation context also be the key to grasping how human practices and work is fundamentally changed by new technology.[3]

As background, the routine capability perspective builds on Brian Arthur's foundational treatise on the nature of technology. While encompassing and enlightening, particularly regarding advances in technology, Arthur's work focuses largely on material devices and their engineering and delves little into routines and device use. My own work extends this view, drawing from Theodore Schatzki's practice theory, to incorporate routines into an overarching perspective in which devices and their use are given a new place in the larger scheme of things.[4]

This new perspective suggests that technology manifests itself in four constitutive and contextual spheres: those of *worlds*, *practices*, *routines*, and *devices*. First, the various worlds in which we live and work are substantially constituted from our human practices. These worlds in turn provide the contexts for the advancement of practices. Second, our various practices are constituted largely from families of routines that provide capabilities. Routines are themselves purposed and developed in the context of human practices. Third, routines and their actions are substantially constituted from devices that provide associated actors with affordances. Devices not embedded in other devices provide affordances only in the context of routines. Figure 6.2 summarizes.[5]

As an everyday example, consider the world of a family's *housework*. Among the practices which constitute it are house cleaning, laundry, cooking, dishwashing, and shopping. Each is constituted from routines making use of devices for carrying it out, such as for shopping, those for fetching groceries by auto, or ordering supplies online for home delivery or pick-up at a store.

Or, consider the world of higher education in colleges and universities. Among the practices that constitute it are teaching, research, administration, IT services, housing, food services, transportation, parking, and building services. Again, each is constituted from routines making use of devices for carrying it out, such as for administration, those for student recruiting and admissions, which we discussed earlier in Chapter 3, and faculty and staff hiring and promotion.

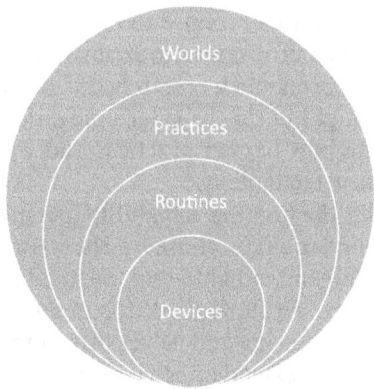

Figure 6.2 Four spheres of technology. Adapted from Swanson (2019).

Explanatory key: Technology manifests itself in four constitutive spheres: those of worlds, practices, routines, and devices. The nesting of the spheres portrays their relatedness. From the outside in, first, the various worlds in which we live and work are substantially constituted from our human practices. At the same time, these worlds provide the contexts for the advancement of practices. Second, similarly, our various practices are constituted largely from families of routines that provide capabilities. Routines are themselves developed in the context of human practices. Third, routines are substantially constituted from devices that provide affordances. Devices not embedded in other devices provide affordances only in the context of routines.

We thus see how devices, or artefacts, are necessarily fused with routines in the larger scheme of things. All four spheres are necessary to an understanding of technology, which, in short, is that devices provide needed affordances, not alone, but as incorporated in routines, which provide needed capabilities, not alone, but as incorporated in practices, which advance or not in their respective worlds.[6]

From the vantage point of this perspective, we also gain new insight into technological change, bringing routines to the foreground of the interpretation. We see that change in routines likely arises in conjunction with change in the devices that afford them as well as with change in the practices that draw upon them. Change originating in any of the spheres is likely to entail change in the related others.

Origins and Modes of Change

But how does change originate from this new perspective? Here, we can draw inspiration from work in evolutionary economics which departs from traditional theory to consider the sources of innovation and technological change, noting that useful knowledge and skills are unevenly distributed in any economy, and that the profit motive stimulates continuous search for better goods and services and means of production. As described by Metcalfe, useful knowledge and technology are fundamentally "restless," and "there are always good reasons to know differently." Where technology is concerned, the economy can't be considered a system in equilibrium, as it is in neoclassical theory. Rather, it must be seen as dynamic, in flux, and as an ongoing problem-generating and problem-solving structure.[7]

Such evolutionary theory seeks to understand how a society or economy learns and advances (or not). Notably, it meshes with organization theory suggesting that relatively invariant *routines* guide behaviors. Paradoxically, routines become the focus of change, notwithstanding their stability, which otherwise works against it. Thus, as nicely summarized by Dosi and Nelson:

> Precisely because there is nothing which guarantees, in general, the optimality of these routines, notional opportunities for the discovery of 'better' ones are always present. Hence, also the permanent scope for search and novelty. Putting it another way, the behavioral foundations of evolutionary theories rest on learning processes involving *imperfect adaptation* and *mistake-ridden discoveries*. This applies equally to the domains of technologies, behaviors and organizational setups.[8]

From the perspective articulated here, it is device-enabled routines and their capabilities that constitute technology itself, which evolves in the context of broader change in human practices. A basic assumption is that in the light of economic and other needs as just described, *humans seek to advance their practices*. When we speak of advancing a human practice, we will mean achieving a more favorable social position for it. This often entails improving upon its economics, social acceptance, or politics, simplifying it or otherwise reducing its costs, for instance, or finding new outlets for it, expanding its presence, or increasing its appeal, making it more enjoyable, or obtaining favored social treatment for it, or improving its reputation. A broad array of purpose gives rise to technological change from the routine capability perspective.

I describe in recent work how change in technology as routine capability evolves through change in the four spheres. Four principal *modes* of change are posited to permeate the spheres: (1) *design*, applied prominently to devices, but also to routines; (2) *execution*, applied to routines, but also to operation of devices; (3) *diffusion*, applied to advancing practices, by spreading routines and devices to a population's members; and (4) *shift*, applied to a world's multiple practices, and how they cohere and compete, opening up (or closing down) opportunities for advancement. Table 6.1 provides an explication. The practice illustrated is business operations and

Table 6.1 Technology Change in Devices and Routines. Adapted from Swanson (2019)

Change mode	Devices	Routines
Design	**Build and test** Example: develop new enterprise software requiring new work routines	**Compose and instruct** Example: develop new work routines for using new enterprise software
Execution	**Operate and maintain** Example: provide fixes and new functionality to enterprise software as necessitated or requested in its use	**Perform and improvise** Example: discover new enterprise software affordances in its use or work around the lack of such affordances
Diffusion	**Distribute and extend** Example: market enterprise software to additional adopters and provide new versions to meet their needs	**Replicate and normalize** Example: employ consultancies to help implement purchased enterprise software following best practices
Shift	**Redomain and reinvent** Example: reconceive and redevelop enterprise software as a service and offer it in the cloud	**Adapt and recreate** Example: develop new support routines for using enterprise software provided as a service

Explanatory key: Technology change in devices and routines occurs through four modes: design, execution, diffusion and shift in practices. Change by each mode takes distinctive forms of devices and routines. *Change by design* is creative in nature. By whatever means, a new device is built and tested. A routine, in contrast, is composed and instruction is offered. *Change by execution* occurs in activation of the technology. A device is operated and maintained. A routine is performed and improvisation is undertaken as needed. *Change by diffusion* occurs with spread of the technology to a wider population of adopters. A device is produced in quantity and distributed and extended in its features. A routine is replicated from place to place, and normalized in its conduct. *Change through shift* occurs as practices ebb and flow, advance and decline, among the practices. A device may be redomained or reinvented to meet changing needs. A routine may be adapted or recreated. In both cases, new design cycles maybe initiated.

the work routines are those which make use of enterprise software. Recent research suggests that achieving routine capability with this ERP technology may involve downsizing and delayering of the organization, decentralizing responsibilities for some tasks while further concentrating control, increasing the range and depth of skills for some jobs while deskilling and routinizing tasks elsewhere, and intensification of work in many of the organization's jobs. While business operations may thus be advanced (if successful) by ERP technology, the ramifications for the newly enabled routines among human participants and their work are seen to be various and complex.[9]

Change in Work and Practices

Clearly, where a firm implements and employs enterprise software, there are likely winners and losers when it comes to individual work. The practice of operations management is presumably advanced, through gaining greater control over the routines that constitute it, embracing the business logic driving the adoption decision, reinforced by the technology's diffusion and institutionalization. Whether the new work routines improve the everyday lives and career prospects of employees is obviously another matter.

Meanwhile, these employees also have housework to attend to at home, as mentioned above and it is useful to reflect on changes here too, in particular where it comes to the advancement of shopping practices. Not all of the routine capability for today's online shopping resides in the home. Rather, it resides among businesses on the Web, as with Amazon's platform, and with payment systems such as those provided by Visa or PayPal or others, and with delivery systems such as those provided by UPS or FedEx or others (evidently soon to use drones). Here we are reminded that where work involves *transacting* among people and organizations, the routine capability that must be built is rather sophisticated and distributed and has no singular business logic. Shoppers are unlikely to change their practices where it makes their lives more difficult. They retain more control over their work in this respect.[10]

It is interesting to speculate whether such control over one's employment work may also gravitate toward a transactional model, as careers are built across shorter terms of employment in what has been termed the "gig economy." Notwithstanding the burdens faced by workers in such an economy, businesses may be challenged to make work more attractive, in particular where they wish to retain professionals committed more to their own practices, than to conventional business rationales. Businesses may need to build their own enterprise more as a platform on which mutually productive and enjoyable work can be undertaken, notwithstanding their corresponding need for operational control of the business as a whole. They may need to attend more to the individual's need to intertwine his or her routines at work and at home. How far this will go remains to be seen. However, that employees will expect that their work environment provides for routine capabilities as supportive and friendly as those that underpin their online shopping at home seems relatively certain.[11]

A broader lesson here is that in looking to the future of work, with new technology, we must look beyond traditional industries and their employment models. We must consider human practices more broadly to gain insights. After all, in a traditional industry such as manufacturing, with automation, there may be less work to be had overall, and information systems will play an important role in bring this about. We may need to look elsewhere. The suggestion here is that to gain needed insight, we should focus closely on everyday human practices, such as shopping, as just touched upon, and the changes they are undergoing with new technologies based in information systems and where this may take us. We expand upon this next.[12]

Consider, for exploratory examples, location technologies, influence technologies, and learning technologies and the routine capabilities they provide to the everyday individual practices of engaging local transportation for oneself, purchasing goods and services, and doing coursework to advance one's education. Each of these everyday practices calls not only upon our routine capabilities as consumers, but also on the routine capabilities of those who organize to serve us in meeting our wants and needs. Ideally, new technologies enable both consumers and providers to advance their respective practices.

Note at the outset that the three technologies chosen are really families of technologies that take various forms as routine capabilities in different settings. We do not explore the technologies in their entirety. We consider only examples of each for brief comparative purposes. Our aim is simply to seed thinking about change and its ramifications from a routine capability perspective.[13]

Location Technology

Consider first location technology as exemplified in arranging local transportation by Uber or Lyft. The routine capabilities provided to both rider and driver are easily understood. The smartphone and its app, in communication with a central dispatching facility, provides the device through which capabilities are built and honed through repeated execution in use. Trips are arranged by an algorithm according to the locations of the parties. While the algorithm has received much attention for its sophistication, the real magic of this technology lies in the value of its routine capabilities to riders and drivers. Much of this value is in the simplification of the ride arrangement, including the payment.[14]

Recent research suggests that routine flexibility is an important source of value to Uber drivers, who expand or contract their engagement over a work day according to the opportunities they have to secure rides and income.[15]

Uber has also expanded its business to include fetching fast-food meals for customers, which involves building new routine capabilities beyond those of providing rides. Restaurants cooperate as participants in building these new capabilities. There is speculation that this may lead to the reduction of on-site dining and the need for servers. Employment loss is a prospect in this eventuality.[16]

On the whole, it is difficult to assess the future for work in local transportation as with Uber. The effects of location technologies are likely to play out in the larger local transportation system, where there are substantial substitution opportunities. Indeed, Uber recognizes this and has recently initiated a program by which a rider can coordinate choices with the local bus and rail systems, for instance.[17]

Influencer Technology

Consider next influencer technology as exemplified in the rise of individual "influencers" who provide consumer guidance on the Web for those who purchase everyday products and services. Such influence work has become a new practice. According to one report, there are now 4.5 million "mom influencers" in the United States, moms who on the web influence other moms in their household purchases.

> They use websites and social media to record seemingly every detail of their lives: the sweater they bought for the fall, their child's favorite new toy, which coffee helps them wake up. What began, for many, as a creative outlet or a way to build community has morphed over the years into big business with a more direct link to brands and companies.

An annual Mom 2.0 conference brings the influencers together. Successful influencers may have as many as a quarter million website visitors per month. Here the devices employed include the camera and tripod, the Web and social media. There is no centralized coordination. However, a vast multi-organizational complex of social data as to posts, friends, tweets, followers, likes, and so on, as well as algorithms for analysis and stimulation of engagement, underpins the influencer practice and its routines. The capabilities built likely differ widely, but influencer work is except for the time demand hopefully complementary to that which moms must undertake at home.[18]

What is remarkable about this influencer work is that much of it is undertaken without employment and that its recognizable value to those who do it comes from being successful at it. The routine capability achieved is the capability to influence itself. Extrinsic rewards can follow, for instance, in paid ad placements on the influencer's website, generated in the larger influence ecosystem. The capacity of this larger ecosystem to engage and influence individuals is largely unbounded and reinforced through its own success. Influence begets more influence. As certain influencer work grows, leveraged by social media, it competes only with the time given to everything else.

Still, for most people, influencer work will probably be a part-time occupation at best. It is difficult to make a living at it. The gains are likely to go to the relatively few who achieve celebrity with it.

Learning Technology

Third, consider learning technology as employed in coursework taken to advance one's education. E-learning in support of business and other college studies has been much discussed and written about, in particular in the distance learning context. Research suggests that it can compete effectively with traditional classroom instruction, although established schools have been slow to embrace it. Student engagement in the e-learning process is found to be a key to its success. Such engagement can be understood as building routine capabilities for both the student and the instructor and support staff.[19]

Not surprisingly, a major attraction of e-learning for the student is the flexibility it allows in achieving routine capabilities. In contrast to fixed scheduled classroom instruction on campus, e-learning can be flexibly woven into other everyday practices and routines that make up the student's life. The flexibility of major interest here is not the flexibility *within* a particular e-learning routine, important though it may be, so much as the *aggregate* flexibility of e-learning routines in support of the broader educational practice.

Less studied has been the appeal of building routine capabilities for the e-learning instructor and support staff. Not surprisingly, resistance to change on the part of those heavily invested in the traditional classroom model has been strong. The virtues and rewards of face-to-face teaching and learning are understandably underscored. But too, it is less obvious that instruction in the e-learning context allows for more aggregate routine flexibility for the provider. Rather, something of the opposite can prevail where the availability of the instructor and staff to field a constant stream of student questions and respond in a timely matter online becomes important to the process.

Still, the future for e-learning work on the part of both student and provider would seem to be robust. The demand for distance learning is substantial and increasing, and the need for accreditation of what is learned is socially fundamental. Most schools should eventually be drawn in, and with the outbreak of Covid-19 and the shut-down of classroom instruction, many already have been, and have developed new routine capabilities. Resistance has been overcome out of necessity. Looking to the future, leading schools can leverage their reputations by offering distance education as part of blended programs, using both campus facilities and e-learning, in attempts to gain the benefits of both.

Insights

Our brief exploration of three technologies as routine capabilities yields several insights regarding the future for work. First, the values of these technologies to those who work with them may be very different, but they must mesh to be successful. Both riders and drivers must benefit from Uber service. Both mom influencers and their followers must find their social media engagement worth their time. Both students and teachers must find e-learning rewarding.

Second, routine flexibility emerges clearly as an important source of value to those who work with technologies. This flexibility is important not just within a routine, but in routine aggregation within or among different work and practices. Technologies are most likely to be embraced where they provide for routine flexibility.

Third, the prospects of new technologies for work and paid employment are by no means the same. Work will always be there to be done, especially in everyday forms such as those explored here. Local transportation will always be a broad need. Influence, sought or not, will always be a feature of life spent with social media. Advancing one's knowledge and skills will increasingly be a lifelong undertaking. But full-time employment and a living wage from work are another matter altogether.

In sum, this essay suggests that when we ask how new technology changes the nature of work, we should look in particular to the routine capabilities achieved in advancing human practices. From this perspective, digitalization doesn't so much "impact" work as such. Rather work itself is changed when people and organizations employ new digital devices to forge new routine capabilities in the interest of advancing certain practices. Work is carried out primarily through routines and what is good or bad about a technology is exposed primarily through routine execution. Studies of work that focus on the routine capabilities that make it productive, attractive and enjoyable or not to the workers who engage in it may thus have much to teach us.[20]

Reflections

Summarizing, we have now arrived at a new place with our essays. I began the book with a traditional take on information systems, as designed to support an organization in its actions. The focus was on IS as *internal* to the enterprise. While the focus incorporated firm-level interactions with suppliers and customers, as well as coordinative interactions within the business, it gave no attention as such to the distinctive human *practices* that an enterprise draws upon and internalizes in the form of various functional units tied together under its umbrella. In this chapter, as we have seen, I have rotated our focus, away from organizations and their interests, toward human practices and their interests. I have argued that technological change is best understood in the context of change in human practices among and even in some ways apart from organizations. I have argued too that the key to understanding this change is to grasp the routine capabilities that advance the practices. This brings a very new and I hope insightful perspective to information systems as technologies.

Organizations come and go, rise and fall, often dramatically, with technological change. Human practices change more slowly, to the extent they prevail in many places, if not everywhere. But they do change profoundly and this speaks more directly to what we gain from technologies such as information systems.

It remains to ask how we might do better with our information systems in the years ahead, given our new understanding of their place in the broader scheme of things. I undertake to answer this question next, in our final chapter.

Notes

1 See Arthur (2009) for a rich exposition on technology from an evolutionary economics perspective. Swanson (2017) explores the application of Arthur's theory to information systems, in particular.

2 See Yoo et al. (2012) on digitalization. Shim et al. (2019) discuss the prospects for the Internet of Things (IoT) in the context of information systems.

3 See Swanson (2019). See too Orlikowski and Scott (2008) on sociomateriality. Becker (2004) provides a review of organizational routines. Feldman (2000) and Feldman and Pentland (2003) examine routines as a source of organizational change. Leonardi and Barley (2008) speak to the challenges to building better theory.

4 Schatzki (2002) provides the important practice theory which I draw upon in my own work (Swanson, 2019).

5 Pozzi et al. (2014) reviews the concept of affordance as it has been applied in the IS literature.

6 Gordon (2016) provides a comprehensive study of U.S. housework practices and how they have changed through new technology since the Civil War.

7 The notion that the economy is "restless" is taken from Metcalfe (2010, p. 160).

8 Quotes are from the review by Dosi and Nelson (1994, pp. 154–155 and p. 159). On routines, see also Nelson and Winter (1982) and Pentland and Feldman (2008).

9 See Hall (2002).

10 It is interesting here to reflect on changes in shopping practices with the outbreak of Covid-19, which has made everyday lives more difficult for so many. Sheltering in place, many consumers have shifted their shopping to online sources such as Amazon that also provide for home delivery. The extent to which this practice shift persists post-Covid remains to be seen. The ramifications for cities and their downtowns and shopping centers are a concern and now being discussed.

11 See Irwin (2016) on the burdens faced by workers in the gig economy.

12 Change in everyday practices, routines, and devices may be associated with profound social change. Arnold (2013) provides an illuminating description of four everyday devices—bicycles, rice mills, sewing machines, and typewriters—and how they were appropriated and assimilated to transform life in India in the late 19th and early 20th centuries. Wang et al. (2020) speak to the present transition from the factory

model of employment to what they term "digital nomadism" in which the individual makes his or her own way in the knowledge economy of the future.

13　Note, in particular, that location technology has many forms and applications. Smartphone apps emit a constant stream of signals enabling its carrier's location to be tracked precisely throughout the day, with worrisome implications for privacy (Warzel and Thompson, 2019). Note too that Influencer technology has widespread forms and applications beyond moms, not only in advertisements and the like, but in politics and cyber warfare.

14　Uber's website offers detailed information on its various services. See, for example, https://www.uber.com/us/en/drive/services/shared-rides/ for a description of the routine followed by drivers in providing shared rides. Conger (2019) reports on the current challenges facing Uber. One of these are proposed laws requiring Uber to treat its contractor drivers as employees. While Uber claims its core business is technology, not rides, note that from the routine capability perspective, the two are inseparable. Cramer and Krueger (2016) provides an analysis of the efficiency of Uber's local ride service in comparison to that of taxis.

15　See Chen et al. (2019).

16　Rogers (2017) includes reflections on Uber and the future of low-wage work.

17　See https://www.uber.com/us/en/ride/transit/.

18　See Krueger (2019).

19　Research on online learning in business schools attracted early attention in the 1990s. Alavi and Leidner (2001) offer an assessment and call for further work. Arbaugh et al. (2009) survey the research on online learning in the different business disciplines. Redpath (2014) argues the case for e-learning, confronting the apparent biases against it. Morgan-Thomas and Dudau (2019) examine the issue of student engagement in e-learning, in particular.

20　See, for example, Boland et al. (2007) on digitalization and the advancement of architecture practice.

References

Alavi, M., and Leidner, D. E. (2001). Research commentary: Technology-mediated learning—A call for greater depth and breadth of research. *Information Systems Research, 12*(1), 1–10.

Arbaugh, J. B., Godfrey, M. R., Johnson, M., Pollack, B. L., Niendorf, B., and Wresch, W. (2009). Research in online and blended learning in the business disciplines: Key findings and possible future directions. *The Internet and Higher Education,* 12(2), 71–87.

Arnold, D. (2013). *Everyday Technology: Machines and the Making of India's Modernity.* Chicago, IL: University of Chicago Press.

Arthur, W. B. (2009). *The Nature of Technology.* New York: Free Press.

Becker, M. C. (2004). Organizational routines: A review of the literature. *Industrial and Corporate Change, 13*(4), 643–678.

Boland, R. J. Jr., Lyytinen, K., and Yoo, Y. (2007). Wakes of innovation in project networks: The case of digital 3-D representations in architecture, engineering, and construction. *Organization Science, 18*(4), 631–647.

Chen, M. K., Chevalier, J. A., Rossi, P. E. and Oehlsen, E. (2019). The value of flexible work: Evidence from Uber drivers. *Journal of Political Economy, 127*(6), 2735–2794.

Conger, K. (2019). Uber fights to get its edge back. *New York Times,* November 1, 2019.

Cramer, J. and Krueger, A. B. (2016). Disruptive change in the taxi business: The case of Uber. *American Economic Review, 106*(5), 177–82.

Dosi, G. and Nelson, R. R. (1994). An introduction to evolutionary theories in economics. *Journal of Evolutionary Economics,* 4, 153–172.

Feldman, M. S. (2000). Organizational routines as a source of continuous change. *Organization Science, 11*(6), 611–629.

Feldman, M. S. and Pentland, B. T. (2003). Reconceptualizing organizational routines as a source of flexibility and change. *Administrative Science Quarterly, 48*(1), 94–118.

Gordon, R. J. (2016). *The Rise and Fall of American Growth.* Princeton, NJ: Princeton University Press.

Hall, R. (2002). Enterprise resource planning systems and organizational change: Transforming work organization? *Strategic Change, 11,* 263–270.

Irwin, N. (2016). With 'gigs" instead of jobs, workers bear new burdens. *New York Times,* March 31, 2016.

Krueger, A. (2019). When mom slams a brand. *New York Times,* November 26, 2019.

Leonardi, P. M. and Barley, S. R. (2008). Materiality and change: Challenges to building better theory about technology and organizing. *Information & Organization, 18,* 159–176.

Metcalfe, J. S. (2010). Technology and economic theory. *Cambridge Journal of Economics, 34,* 153–171.

Morgan-Thomas, A., and Dudau, A. (2019). Of possums, hogs and horses: Capturing duality of student engagement in eLearning. *Academy of Management Learning & Education, 18*(4), 564–580.

Nelson, R. R. and Winter, S. G. (1982). *An Evolutionary Theory of Economic Change.* Cambridge, MA: Harvard University Press.

Orlikowski, W. J. and Scott, S. V. (2008). Sociomateriality: Challenging the separation of technology, work, and organization. *Academy of Management Annals, 2*(1), 433–474.

Pentland, B. T. and Feldman, M. S. (2008). Designing routines: On the folly of designing artifacts, while hoping for patterns of action. *Information and Organization, 18,* 235–250.

Pozzi, G., Pigni, F. and Vitari, C. (2014). Affordance theory in the IS discipline: A review and synthesis of the literature. *Proc. Americas Conference on Information Systems 2014.*

Redpath, L. (2012). Confronting the bias against on-line learning in management education. *Academy of Management Learning & Education*, *11*(1), 125–140.

Rogers, B. (2017). The social costs of Uber. *University Chicago Law Review Online*, *82*(1), 6.

Schatzki, T. (2002). *The Site of the Social*. University Park: Pennsylvania State University Press.

Shim, J. P., Avital, M., Dennis, A. R., Rossi, M., Sørensen, C. and French, A. (2019). The transformative effect of the internet of things on business and society. *Communications of the Association for Information Systems*, *44*(1), 5.

Swanson, E. B. (2017). Theorizing information systems as evolving technology. *Communications of the Association for Information Systems*, *41*, 1.

Swanson, E. B. (2019). Technology as routine capability. *MIS Quarterly*, *43*(3), 1007–1024.

Wang, B., Schlagwein, D., Cecez-Kecmanovic, D. and Cahalane, M. C. (2020). Beyond the factory paradigm: Digital nomadism and the digital future (s) of knowledge work post-COVID-19. *Journal of the Association for Information Systems*, *21*(6), 10.

Warzel, C. and Thompson, S. A. (2019). How your phone betrays democracy. *New York Times*, December 21, 2019.

Yoo, Y., Boland, R. J. Jr., Lyytinen, K. and Majchrzak, A. (2012). Organizing for innovation in the digitized world. *Organization Science*, *23*(5), 1398–1408.

7 How Can Information Systems Make a Better World?

On August 19, 2019, a Workshop on Information Systems Research and Development (WISRD) was held in Santa Barbara, California, where I now live. The theme of the workshop was: How Can Information Systems Make the World a Better Place? A dozen or so former students and colleagues led a series of discussions on current topics of interest. It was striking to me that our different expressed concerns around the future for information systems addressed enduring human aspects that have long challenged the systems approach taken in our work.[1]

Among the issues discussed were these:

• Technology and the displacement of human work
• Technology and surveillance
• Technology and political warfare
• Technology and evil
• Technology and managerialism

Many years ago, my mentor C. West Churchman published an under-recognized book, *The Systems Approach and its Enemies*, that addressed the challenges to the systems approach, understood as a form of rational planning and design, embodied in practices such as operations research and systems analysis. The four "enemies" of the practical reason underlying the systems approach were identified by Churchman as politics, morality, religion, and aesthetics. All have well known philosophical roots and speak to human action. As interpreted here, politics manifests itself in forming polis, coming together as "us" to act collectively. Morality is the spirit and compass that guides action toward what is right and "good." Religion encapsulates what is held to be "sacred" in the context of action. Aesthetics is that which gives life to action and makes it radiant or "beautiful" for us.[2]

DOI: 10.4324/9781003252344-7

Just as the systems approach devotes itself to the pursuit of truth over falsity in applied contexts, each of the enemies incorporates its own polar opposites, creating tensions in our actions: politics distinguishes between us and "them"; morality between the good and the "bad"; religion between the sacred and the "profane"; aesthetics between the beautiful and the "ugly." Accordingly, beyond commitment to the truth, we should not be surprised to find these enemies and their tensions reflected in the issues which concern us about the future for information systems.

Tensions abound in the short list of issues above. For instance, with the displacement of human work comes the draining of its rewards, both material and intrinsic. Whether this is good or bad for us is a contentious judgment that will be made regardless of any analyst's tally of its costs and benefits. As another example, with widespread digital surveillance comes a certain "chilling effect" on private autonomous action, as well as manipulation of our actions through intrusive knowledge of our everyday behavior. What are the ethics of such surveillance? Under what circumstances might we find surveillance not only wrong, but ugly?[3]

Consider too the rising use of information systems in political warfare and the apparent breakdown of national and social polis into a multiplicity of aggrieved us versus them, the *other* that we would wish away, such as the recent immigrant, or those we hold in contempt, such as the too well educated big-city elite. And with such warfare, the attack on reasoned truth by well-armed disputants to it, as with persistent fortified climate change denial. And the mounting of dis-informational attacks on journalists and news media that as "enemies of the people" stubbornly pursue stories that we do not wish to hear told.

Today, it would seem that we have arrived at the end of a certain Age of Innocence for information systems, if ever there was one. The systems approach we have long taken seems to have now run aground among its enemies. If in the beginning, information systems concerned itself with conveying simple truths such as the number of items in an inventory, today through our efforts over decades, we find our systems everywhere and entangled in all human action. There is no getting away from it. Whatever goes right or wrong, we seem to have a hand in it. As we now consider the future for information systems, we must frame it in the broader human context where in seeking to convey truths that matter greatly to us all, we do this with due regard for the "enemies" that make us fully human.

In this last chapter, I seek some insights into the future for information systems, and attempt to provide some guidance as we look ahead. Today, it is fashionable to assert that we as practitioners and researchers have moved on from traditional information systems to a new era of digital innovation. Most eyes seem fixed on the bright (or not) horizon that lies before us. The theme of this book is that we have not moved on from information systems at all. Rather, we have continued to build out systems in new forms to underpin all of our digital efforts. Moreover, information systems are now central to everyday life facilitated by the digital, though they operate mostly out of sight. Looking ahead, we need to keep this accomplishment and consequent responsibility very much in mind.

Information Systems Revisited

Before addressing the future for information systems, it is necessary to revisit here some of its basics and the insights drawn in earlier discussions. This will provide something of a grounding for what we venture in looking ahead.

First, recall that we view information systems as computer-based systems that provide information to guide organizational actions. Our perspective is an organizational one. And it concerns itself with the actions of organizations. While the actions of people as users are also important, whether as workers or managers, or as customers, or as citizens, most systems serve the organization first of all. Measures of performance are typically organizational ones.

Second, in elaborating on why organizations have information systems, we have argued that it is to facilitate enterprise-level actions and interactions, subunit-level actions and interactions, and system-level actions and interactions. By design, most information systems aim to coordinate multi-level interactions among multiple players, within and across organizations. This focus on interactions is fundamental to our perspective.

Third, in addressing what systems provide in coordinating interactions, we have defined organizational information as purported facts given and taken, and inferences drawn and established by participants in an open interaction network. Participants here include both the machines that draw upon organizational data and code, and users and others that draw upon their knowledge in their network interactions. Information has a temporal quality and is created and maintained in the interactions.

Fourth, we have explored the everyday interactions among human users of information systems and the internet, and have identified four distinct forms: informational, cooperational, transactional, and social. We have seen that these different forms extend the interaction context for information systems beyond their traditional foundations and purposes, especially with the expansive use of social media, presenting both opportunities and challenges to the enterprises that house and deploy them.

Fifth, we have sketched the history of information systems as a practical endeavor since its origins and found it to be anchored fundamentally in facilitating organizational transactions. We have shown that the transactional path has taken us from accounting systems to enterprise systems to retail automation and to electronic commerce. And we have shown that information systems have now come to underpin everyday transactions around the world.

Sixth, we have explored the foundations of information systems not just in organizations, but in human practices more broadly. We have examined technological change in advancing human practices and identified four modes of change: design, execution, diffusion, and shift in practices. We have suggested that to understand the future for information systems, we should pay close attention to the longer-term ramifications for everyday human practices, quite apart from traditional organizational rationales for their deployment as tools of coordination within and among enterprises.

In sum, from earlier discussions, we observe that the foundations for information systems have shifted since their inception from their original, rather narrow organizational purposes. In particular, with the rise of the internet and new enterprises such as Google and Facebook that are built on internet and public web foundations that facilitate everyday interactions by people in roles beyond organizational confines, we see that systems must now be collectively built to facilitate interactions across an archipelago of players, more than to serve an organizational island of one.

We find that the future for information systems entails building important social foundations beyond the traditional organizational ones. But what should these foundations be and how should we go about establishing them?

Looking Ahead

In what follows, I venture several guidelines we might follow as we engage with information systems, oriented toward making the world a better place, drawing from what we have learned in this book thus far. Each guideline reflects a commitment to the practical truths to be gained from information systems, but is fashioned with a wary eye on the enemies of the systems approach as discussed above.

Among the four enemies, it is politics that has been particularly incited in its polarities by new technologies built with information systems. Social media, in particular, in extending their reach beyond the local community (college campus, in the case of Facebook) in which they were first envisioned, have given new interpretation to the old truism that "all politics is local." One is tempted to say that on the Internet and Web, all politics can with sufficient incitement be extended to the global (or at least to new locals beyond the singular physical place). Today, our politics seems heavily invested in the furthering more than the tampering of this incitement, with the consequence that its presence as an enemy of the systems approach is pervasive.

In the case of morality, perhaps the greatest challenge pertains to the global economic system and its pursuit of growth, with consequences for the distribution of wealth among and within peoples, as well as the plundering of natural resources to the point of endangering the future of life on earth. With global warming and consequent climate change now unavoidably upon us, and as systems of all kinds fail us, it is likely that "business as usual" will increasingly be judged morally unacceptable and that information systems that simply sustain this doomed order will be seen as part of the problem, rather than necessary to the social solution. Or so the moral challenge is likely to be made by this enemy of the systems approach.

And so, in looking ahead to the future for information systems, we must grapple, in particular, with the challenges associated with morality and politics as enemies of the systems approach. We must keep an eye too on religion, for instance, where the earth is held to be sacred, or of no important matter and undeserving of being held in awe. Such religious tensions may give added impulse to the oppositional forces generated by both morality and politics as enemies that undermine practical reasoning in terms of what is true and false. Similarly, aesthetics will play its important role in inspiring, but also critiquing our ongoing systems endeavors.

With the enemies in mind, then, I venture five provisional guidelines for an information systems approach that might help make the world a better place. I make no attempt at being comprehensive with these guidelines. Rather, I try to get at something very basic and important with each of them. I offer them up in somewhat fragmentary form for further consideration and discussion. I note where the enemies of the systems approach lurk.

Nurture Human Agency

While information systems support organizations in their actions, they are simultaneously entwined in human practices, both work practices and other practices apart from work, as we have seen. Failure to attend to this can result in systems that achieve organizational goals at unacknowledged human expense.

Most information systems have important, sometimes even profound, ramifications for human practices and their routines. Not infrequently, organizational imperatives result in systems that not just inform but also constrain human action in routines. The human *sense of agency*, that one's actions are consequential and that one is responsible for them, can be subtly undermined, where it needs instead to be nurtured.

The physician and writer Atul Gawande describes the challenges and changes doctors face in their interactions with patients and electronic medical records systems with which they must work directly. He explains "why doctors hate their computers" in terms of the loss of *quality* in human and machine interactions overall. Research finds that physicians in office visits now spend as much time interacting with the system as they do with the patient, and yet more time with the system outside the visits and outside normal working hours. Burn-out is rampant. In his own case, Gawande laments that, "...I've come to feel that a system that promised to increase my mastery over my work has, instead, increased my work's mastery over me."[4]

Gawande's remarks would come as no surprise to scholars who have studied the fortunes of enterprise systems in organizations since the 1990s. As we have earlier discussed, these systems are by design notoriously constraining in human interactions that engage them. Their logic and data essentially rule the enterprise's business processes. Their implementation and assimilation is often an ongoing struggle.

One can see how aesthetics in work plays a subtle underlying role here too. Information systems can make it difficult for people to find beauty and joy in their routines, in particular when they are tethered to their devices rather than free to employ them to their own reward and pleasure. The challenge is to find new ways for people to achieve the joys of mastering their work in interactions with information systems.

As a general principle, information systems should not be designed to achieve organizational goals in *ignorance* of their ramifications for human practices and their routines. Rather, goals should acknowledge the need to nurture human agency in a practice's routines. Should goals thus include a promise to increase "mastery over work," as with the physician example, designs should also aim to make good on it. Workers should be empowered in their interactions, by whatever means. Systems should be designed to serve practices, not to put humans into organizational subservience.

In the long run, the world is likely to be a better place where we focus on our human practices, not just on organizational goals such as cost reduction, or increased market share, which speak to the viability of a particular firm or organizational form, more than on the human practices engaged in around this form.

Promote Transactional Equity

While information systems support an organization's interactions with other organizations and with people, these systems by their nature largely privilege organizations with both power and information over people. These asymmetries have long been with us around transactions between buyers and sellers in product and labor markets. Thus, where a firm sells its products to consumers, it knows more about its product than does the consumer, and, increasingly, with its systems, may know a great deal about the consumer too. The consumer's knowledge of the firm and its product draws from the firm's public reputation, more than from specifics related to the transaction. And so a firm as a seller has a certain advantage over the individual person as a buyer, in the absence of structures, such as warranties, that provide for transactional equity. Let the buyer beware.[5]

In labor markets, the firm is confronted with the reverse asymmetry, where the job applicant presumably knows more about herself as a prospective employee, than does the firm. Hence, the classic request that the applicant submit letters of recommendation. However, today, the firm can also access a variety of systems to learn more about the applicant and her past and unwise Facebook posts, including those she had herself long forgotten. To the extent these systems are unreliable and lack integrity, and the job applicant does not have the same access to their information, as does the firm, transactional equity is undermined at the individual's expense. Let the seller beware.

In both consumer and labor markets, then, individual people are increasingly disadvantaged in their transactions with organizations in terms of both information and power. Inequities follow. The advent of the era of Big Data has exacerbated the problem, in particular, in the United States, as individual actions are tracked and aggregated into personal data records sold by data brokers in largely unregulated markets.[6]

As a general principle, parties to a prospective transaction should be equally informed as to its terms and conditions and organizational data and business logic pertinent to engaging in the transaction should be shared. And to the extent that such equity cannot be practically achieved, the disadvantaged party should be protected in the law. For example, organizations might better be held liable for damages resulting from the use of incorrect data or faulty business logic in transacting with people in both consumer and labor markets. Today, however, the law arguably favors the organization in its contracts. Business firms increasingly write waivers of rights into their contracts and force arbitration in dispute resolution.[7]

Firms are generally more empowered than people in contracts, because firms write the contracts. Boilerplate legalese protects them. The problem is exacerbated online where consumers must agree to terms of service in engaging a business website. A New York Times editorial has called for an end to terms of service, and for new laws that provide basic guarantees of privacy, concluding "We don't need to live in a world governed by [business service] terms and conditions, propped up by the legal fiction of consent."[8]

Transactional equity might also be served through designed transparencies in the organization's information systems, for example, in the form of embedded meta-data, such that they may be more easily queried by those that work with them and audited by third parties and governmental entities for their fairness. For instance, the sources of a system's data and code might be clearly identified down to a detailed item level.

In the long run, the world is likely to be a better place where social and organizational structures are devised and put in place to promote equity in everyone's everyday transactions.

Cultivate Information Ethics

While information systems can usefully inform an organization in its actions, they can also violate certain information ethics in doing so. They may violate norms of personal privacy or make unauthorized use of code and data owned by others. They may serve to *misinform* others in the organization's interactions. They can do this in various ways, for instance, by simply withholding information that might be shared, or by conveying information that is incomplete or less than accurate.[9]

Information systems can also be used to propagandize more than inform. They can be used to spread outright falsehoods or to undermine unwelcome truths, or to advance unsupported conspiracy theories, or to deploy disinformation in political warfare. Today, stories of the malevolent and warring use of information, derived or not from factual data, are common in the news. From one recent study, the regimes of some 70 countries have engaged in disinformation campaigns in perpetuating their rule.[10]

While the ethical issues associated with information systems are many, the concern here is with the ethics associated with the information itself, that is, the purported facts given and taken, and the inferences drawn and established by participants in an open interaction network. The concern is with issues as to the truth and accuracy of factual claims, the logic and reasoning in inferences, agreements and respect for privacy and confidentiality, security and governance, and access and participation in the network. It is here that our ethics need to be cultivated, in particular.[11]

The information ethics associated with social network platforms present significant challenges at this particular moment. Misuse of these platforms is rampant. What responsibilities do these platforms have for their content and how can they best be met? As one example, Facebook announced that it would not monitor the content of political ads it accepts, places and promulgates in advance of the U.S. 2020 presidential election. Whether these ads are worthy or not and communicate truths or falsities would not be Facebook's concern. This policy was said to be in the interest of free political speech. Predictably, it came under fire.[12]

Few issues have received as much attention as that of information privacy, understood as the extent to which an individual can control information about him- or herself. The End of Privacy in the digital age has been broadly announced and widely lamented. But practically, privacy is not something which one has entirely or not. Rather, the issue is more that of confidentiality of information selectively shared, for instance in our business transactions, or in our health care, or in our filed taxes, or online among friends. The issue is how to best attend to information confidentialities in an open interaction network.[13]

As a general principle, information systems should embrace and cultivate specific information ethics in their design and use. These specifics would be expected to vary among systems, according to their nature. For instance, systems that gather and maintain surveillance data might incorporate retention rules judged to be reasonable under their circumstances. Recognizing that morality as an enemy of the systems approach might challenge this reasoning, whether these rules were good or bad ones would be a subject of ongoing open discussion.

Information systems should also be subject to professional audits that embrace and address information ethics as part of their practice.

In the long run, the world is likely to be a better place where we cultivate information ethics in our information systems and monitor their performance accordingly.

Inform the Social Good

While information systems are designed to inform the organizational good, they are not typically designed to inform the social good. By the social good, I mean here the good, less the bad, generated for others with whom the singular organization interacts or not. In economic terms, this good pertains to both the internalities and externalities of the interactions among all participants. To inform the social good is to enable us socially to comprehend and collectively act on it, a considerable challenge.

The problem for the private organization such as a business firm is that it has no authority to judge the social good or bad for which it is responsible. Rather, it must hear from others affected and respond as appropriate in the socio-political and governance context in which it operates. Authority may reside in governmental oversight and regulation. These public organizations have their own information systems to help guide their actions.

Thus, in the United States, to inform the social good, we look to the work of Federal agencies such as the Food and Drug Administration (FDA), the Environmental Protection Agency (EPA), and the Occupational Safety and Health Administration (OSHA), among others. We also look to state and local agencies, and to non-governmental organizations (NGOs) established to serve the public interest.

The ways in which the social good is informed are many. An illuminating case is the introduction of restaurant hygiene quality grade cards by the Los Angeles County Department of Public Health in 1998, where inspection results came to be posted prominently in windows for all to see. Research found that not only were customers better served by this information, but that hygiene quality in facilities was also improved (with less hospitalization from food poisoning). From our perspective, we note that the information provided was also the product of three-way interactions between the regulatory agency, the restaurant(s), and the customer(s). More broadly, informing the social good likely requires more complex interactions than in everyday exchanges between private parties, and supporting systems must be designed with this in mind (in weighing not only informational benefits, but costs to all parties, for instance).[14]

As a general principle, information systems should be designed to inform the social good rather than to simply presume it.

In the long run, the world is likely to be a better place where governmental and other public organizations devise and make use of information systems that specifically aim to inform the social good.

Attend to Material Consequence

While information systems are designed to inform organizational actions, they are often not designed to monitor the consequences of these actions beyond organizational boundaries, apart from direct concerns such as marketing to customers. Especially where consequences are *externalized*, as with much industrial waste disposal, there may in fact be a disincentive to monitoring, as the information gained may speak primarily to costs imposed on others. It may be thought better to await complaints from these others and deal with the consequences then.

Indeed, when human practices are adversely affected by organizational actions, those who suffer the consequences are not likely to remain quiet, once they know about it. When certain practices are advanced at the expense of others, open contention may follow. But a focus on the consequences for practices alone does not suffice.

The problem is that there is material consequence to our information systems as technology beyond the advancement of our human practices, and it has long-term implications for the global world we live in. By material consequence, I mean that which has no social component as such, but rather exists within the physical and biological world within which our social world is constructed, a larger ever-changing world necessary to us and affected by us, but without particular regard for us. I am referring to material consequence on the material earth.[15]

As a general principle, certain of our practices should be devoted to monitoring and attending to the material consequences of our actions. Much of what we need to know must come from our science endeavors. Climate science, for instance, has advanced in part through monitoring exemplified by the direct measurement of carbon dioxide in the earth's atmosphere from the top of Mauna Loa in Hawaii since 1958. That the surface of the earth should warm from increases in atmospheric CO_2 was known for almost two centuries. Historical thermometer readings indicated that the earth had warmed by a degree Fahrenheit since the beginning of the industrial revolution. Research confirms that fossil fuel burning and other human activities explain the global warming which continues today. The challenge is to translate this material consequence into terms that we understand and accept and will choose to act upon.[16]

Meeting this challenge has proven to be much more difficult than might be expected, as it requires political action and has inflamed polis, especially in the United States, in part because our institutions call for us to inform and persuade ourselves through social speech costly to exercise and in constant need of funding. And so, for instance, the American Association for the Advancement of Science (AAAS), to inform the public of the findings of climate science, reaches out to its membership to make extra contributions toward its efforts. At the same time, organizations that oppose actions to combat climate change marshal their resources to voice denial of the scientific consensus, often through non-mainstream media. The information generated in these interactions is not necessarily guided by the facts and the science behind them. Politics proves to be an able enemy of the systems approach.[17]

In the long run, almost needless to say, the world is likely to be a better place where we attend to material consequence so that the practices we engage in can be adapted in the interest of life on earth.

Conclusion

In this short book, I have tried to bring together some thoughts on what information systems are all about and where we have arrived with them and the future we face. I am not alone in this, of course. The field of information systems scholars is nothing if not introspective with regard to its accomplishments.

Yet it remains difficult to communicate with the broader academic and practitioner communities on the importance of the IS field. Many will recall the infamous claim by Nicholas Carr that "IT doesn't matter," and how academic leaders in IS rose to the occasion to dispute the claim and attempt to put it in its place. Having witnessed the spreading commoditization of IT, Carr had concluded that the strategic game with it was surely over. Management could move on to other things. Like what?[18]

More recently, but a few years back, I attended a lecture by a distinguished management scholar in the field of operations and logistics. As I recall his talk, it addressed four interesting and distinctive breakthroughs achieved with new technologies in supply chains. Three of these came from new forms and extensions of information systems, in my interpretation. Yet the scholar, a modeler, mentioned IS not at all. He did say in passing that "IT was just an enabler" in the cases he described.

Of course, from the perspective of this book all technologies are enablers in that they provide for routine capabilities in practices. This is what is most interesting about technologies, not what is least interesting.

At this writing of the book, my colleagues Richard Baskerville, Michael Myers, and Youngjin Yoo have just authored a new article giving a fresh take on where we are now going with IS. From their perspective, it is now "digital first" as the digital world comes to increasingly dominate the physical world, as they describe it. The newly emerging research streams in the IS scope are seen to be these: computed human experiences, digital technologies shaping our world, digital infrastructures and the entire digital ecosystem, adapting society to information systems, and human values.[19]

No doubt, as information systems continue to unfold as a field of study and practice in the years ahead, we clearly have much to look forward to!

Notes

1 The organizers of WISRD 2019 were Cynthia Beath, Mary Culnan, and Ping Wang, former students to whom I am grateful for putting this special event together. Thanks too to the attendees whose contributions made the occasion so engaging and enjoyable. A copy of the program is available on request.

2 This is clearly a simple interpretation, perhaps overly so. Churchman (1979) uses the word *enemy* to "connote this immense land of social systems that has remained largely unexplored by 'hard' systems analysts"

(p. xi). In his terms, it is unexplored because it eludes and even opposes the rationality and logic inherent to the systems approach and its models. Of course, the territory represented by the four enemies has been extensively addressed by philosophers since the time of the Greeks. Gardner (2011) reexamines the three classical virtues of truth, beauty, and goodness in today's context. Note that the pursuit of truth is the province of the systems approach, while the pursuits of beauty and goodness are provinces of two of the four enemies. The term "enemy" should not be taken pejoratively.

3 See especially Brynjolfsson and McAfee (2014) on the displacement of work. See Zuboff (2015) and Clarke (2019) for trenchant critiques of the surveillance economy.

4 See Gawande (2018).

5 The classic contribution to the economics of information asymmetries is that of Akerlof (1970) on the market for lemons.

6 See Martin (2015) for a useful examination of the Big Data industry and its supply chain.

7 See Silver-Greenberg and Gebeloff (2015).

8 See https://www.nytimes.com/2019/02/02/opinion/internet-facebook-google-consent.html. For background on legal boilerplate, see Boardman (2005) and Hillman (2005).

9 Mason (1986) identifies four ethical issues associated with the information age: privacy, accuracy, property, and accessibility.

10 See Alba and Satariano (2019).

11 Our notion of information ethics tracks closely with the discourse ethics discussed in Mingers and Walsham (2010), which is also a good source for exploring the philosophy of ethics more broadly. For another, more elaborate and ambitious theory of information ethics, see Floridi (1999). Introna (2002) speaks to the challenges of information ethics in an age in which our interactions are increasingly with and through our machines, rather than face-to-face.

12 See Kang and Isaac (2019).

13 See Rotenberg et al. (2015) for a useful collection of essays on privacy. Richards and King (2013, 2014) argue that "privacy" is best understood in terms of rules for sharing information that remains confidential between parties. Confidentiality relies on trust and promises made in the context of relationships. Legal mechanisms to enforce confidentialities are challenging to achieve in the era of Big Data. Hartzog (2011) proposes a "chain-link" confidentiality regime to contractually link obligations to protect information which moves "downstream" among parties.

14 See Jin and Leslie (2003).

15 See Swanson (2018). Slides available from the author. When I say that our physical and biological world "has no particular regard for us" I mean that it doesn't favor us in any way, although our actions sometimes suggest that we believe otherwise or that this is unimportant. Clearly, not everyone subscribes to my understanding. Religion as "enemy" plays an important role.

16 Mann (2012) documents the contention that climate science has faced in winning social acceptance.

17 See the AAAS website on climate science at: https://whatweknow.aaas. org.
18 See Carr (2003). McFarlan and Nolan (2003), among many others, provided a rebuttal.
19 See Baskerville et al. (2020).

References

Akerlof, G. A. (1970). The market for "lemons": Quality uncertainty and the market mechanism. *Quarterly Journal of Economics, 84*(3), 488–500.

Alba, D. and Satariano, A. (2019). At least 70 countries have had disinformation campaigns, study finds. *The New York Times*, September 26, 2019.

Baskerville, R., Myers, M. and Yoo, Y. (2020). Digital first: The ontological reversal and new challenges for IS. *MIS Quarterly, 44*(2), 509–523.

Boardman, M. E. (2005). Contra proferentem: The allure of ambiguous boilerplate. *Michigan Law Revue, 104*(5), 1105–1128.

Brynjolfsson, E. and McAfee, A. (2014). *The Second Machine Age*. New York: W. W. Norton.

Carr, N. G. (2003). IT doesn't matter. *Harvard Business Review, 81*(5), 4–11.

Churchman, C. W. (1979). *The Systems Approach and its Enemies*. New York: Basic Books.

Clarke, R. (2019). Risks inherent in the digital surveillance economy: A research agenda. *Journal of Information Technology, 34*(1), 59–80.

Floridi, L. (1999). Information ethics: On the philosophical foundation of computer ethics. *Ethics and Information Technology, 1*(1), 33–52.

Gardner, H. (2011). *Truth, Beauty, and Goodness Reframed*. New York: Basic Books.

Gawande, A. (2018). Why doctors hate their computers. *The New Yorker*, November 12, 2018.

Hartzog, W. (2011). Chain-link confidentiality. *Georgia Law Review, 46*, 657–704.

Hillman, R. A. (2005). Online boilerplate: Would mandatory website disclosure of e-standard terms backfire? *Michigan Law Review, 104*, 837–856.

Introna, L. D. (2002). The (im) possibility of ethics in the information age. *Information and Organization, 12*(2), 71–84.

Jin, G. Z. and Leslie, P. (2003). The effect of information on product quality: Evidence from restaurant hygiene grade cards. *The Quarterly Journal of Economics, 118*(2), 409–451.

Kang, C. and Isaac, M. (2019). Defiant Zuckerberg says Facebook won't police political speech. *The New York Times*, October 17, 2019.

Mann, M. E. (2012). *The Hockey Stick and the Climate Wars*. New York: Columbia University Press.

Martin, K. E. (2015). Ethical issues in the big data industry. *MIS Quarterly Executive, 14*(2), 67–85.

Mason, R. O. (1986). Four ethical issues of the information age. *MIS Quarterly*, *10*(1), 5–12.

McFarlan, F. W. and Nolan, R. L. (2003). Why IT does matter. *Harvard Business Review Online*. At https://hbswk.hbs.edu/item/why-it-does-matter.

Mingers, J. and Walsham, G. (2010). Toward ethical information systems: the contribution of discourse ethics. *MIS Quarterly*, *34*(4), 833–854.

Richards, N. M. and King, J. H. (2013). Three paradoxes of big data. *Stanford Law Review Online*, *66*, 41.

Richards, N. M. and King, J. H. (2014). Big data ethics. *Wake Forest Law Review*, *49*, 393.

Rotenberg, M., Horwitz, J. and Scott, J., Eds. (2015). *Privacy in the Modern Age: The Search for Solutions*. New York: The New Press.

Silver-Greenberg, J. and Gebeloff, R. (2015). Arbitration everywhere, stacking the deck of justice. *The New York Times*, October 31, 2015.

Swanson, E. B. (2018). Material consequence. *Workshop on Technology Matters and Matters of Technology*, Portsmouth, UK, June 25, 2018.

Zuboff, S. (2015). Big other: Surveillance capitalism and the prospects of an information civilization. *Journal of Information Technology*, *30*(1), 75–89.

Index

Note: **Bold** page numbers refer to tables; *italic* page numbers refer to figures and page numbers followed by "n" denote endnotes.